Mathematics

in a Sample of Cultures

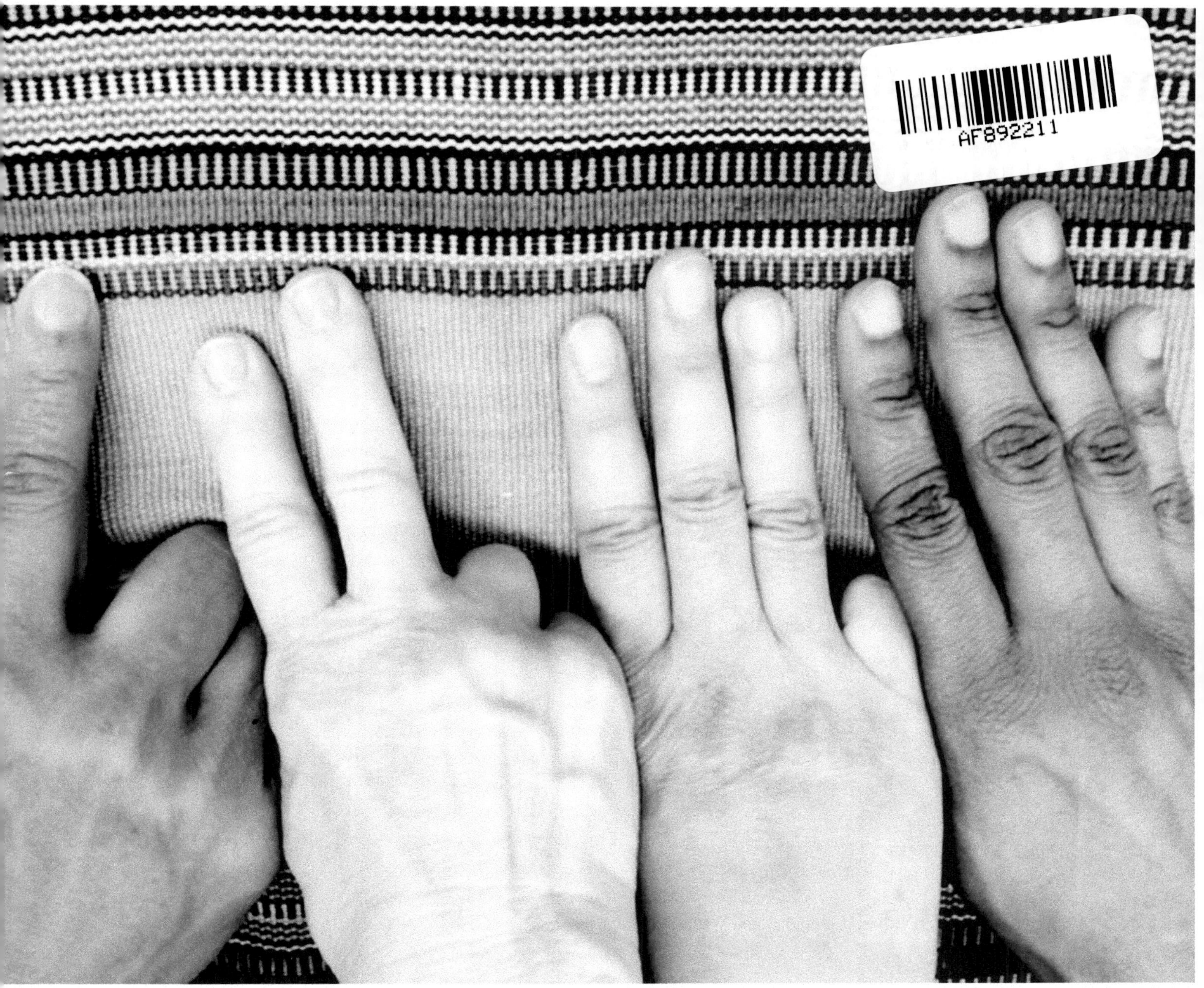

Ximena Catepillán
Millersville University

Waclaw Szymanski
West Chester University

Kendall Hunt
publishing company

Kendall Hunt
publishing company

www.kendallhunt.com
Send all inquiries to:
4050 Westmark Drive
Dubuque, IA 52004-1840

ISBN 978-1-5249-2964-

Printed in the United States of America

Contents

Preface by Authors

A group of hunters, having pitched camp, set out to go bear hunting. They walk one mile due south, then one mile due east, at which point they sight a bear. Bagging their game, they return to camp and figure out that altogether they have traveled three miles. What was the color of the bear?

American riddle

Mathematics can be found at the basis of most human civilizations. It usually begins with the need for counting, which leads to the introduction of natural numbers: 1, 2, 3, 4, ... in the Hindu-Arabic notation.

One of the biggest discoveries in the development of the notation in number systems was the introduction of the symbol for the number zero, which symbolizes something like nothing. It, almost certainly, did not come from counting, because counting nothing is quite contrary to everyday common sense. In the European civilizations, the concept and symbol of zero were originally attributed to the Arabs, but later it was established that the Arabs learned it from India, where it was invented several thousand years earlier.

This information was prominently displayed on the main poster inviting mathematicians to the 2010 International Congress of Mathematicians in Hyderabad, India.

It is amazing to find that the Maya who lived in Mesoamerica and, most likely, did not have any contact with European civilization until the conquista in 1519, or, even less likely, civilization of India, had their own symbol, in fact, many symbols for zero.

The symbol for zero makes possible the positional place value system for expressing numbers in writing. In modern language, it is described as writing numbers relative to a given base in terms of consecutive powers of the base. For example:

-If the base is 10 (decimal system) ten symbols—digits are used: 0, 1, 2, 3, 4, 5, 6, 7, 8, 9 and

$$203 = 203_{10} = 2 \times 10^2 + 0 \times 10^1 + 3 \times 10^0.$$

Here, the subscript 10 indicates the base 10.

The word "digit" comes from *digitus* which means finger in Latin. Now, it is clear where the decimal system came from.

-If the base is 2 (binary system, which is the foundation of all computer technology) two symbols—digits are used: 0, 1 and

$$203_{10} = 1 \times 2^7 + 1 \times 2^6 + 0 \times 2^5 + 0 \times 2^4 + 1 \times 2^3 + 0 \times 2^2 + 1 \times 2^1 + 1 \times 2^0 = 11001011_2.$$

Here, the subscript 2 indicates the base 2.

To appreciate the importance of the concept and symbol for zero just look at the Roman numerals. Romans did not have any symbol for zero and, as a result, the Roman numerals have no positional place value form. For example, 203_{10} is expressed in Roman numerals as CCIII, where C stands for *centum* which is one hundred in Latin (cf. century or cent) and I is one. CCIII reads $100 + 100 + 1 + 1 + 1$. There is no positional place value base here. When the system of writing numbers does not have positional place value, calculations are much harder than when it has one.

Somehow naturally comes a need for addition—the most basic arithmetic operation. Subtraction seems to come next, but subtracting only smaller numbers from greater, in order to remain within natural numbers. Multiplication follows in the form of repeated addition. Division appears to be the most sophisticated operation and comes mainly in the form of division with remainder, again, to stay within the world of natural numbers.

In this book, we describe some basic Mathematics in some non-European cultures.

We begin with the Vedic culture present in India for at least 6,000 years. Even though India is, certainly, a non-European country, the foundations of the entire European Mathematics are in India. Mathematics of India was transmitted to Europe through the Arabs. Hence, the often used term Hindu-Arabic.

In Vedic Mathematics, we find the first known positional place value system of writing, which is decimal, exactly like the one we currently use. The classical language of the Vedic culture is called Sanskrit, the system of writing is called devanagari, and in the system the numerals resembles the one we use presently, for example, the symbol for zero is 0, the symbols for two and three resemble 2 and 3, respectively.

Then, we discuss the Inca civilization of parts of South America, which developed their own positional place value decimal system. Unfortunately, there is no original Inca record of their Mathematics, because they did not have any lasting writing system. Most of we know about it comes from the records of the Spanish chroniclers during the conquest of the Inca empire that began in 1521.

Afterward, we visit the Mesoamerican civilization of the Maya. Here, Mathematics gets a bit more complicated, because the number writing system, even though still positional place value, usually uses the base of 20—thus it is called vigesimal—the explanation of this name is in that chapter. It uses, certainly, different digit symbols, from the Hindu-Arabic system, but it is not dissimilar in basic principles.

As an important application, we describe calculations related to the Maya calendars. Unlike Inca, the Maya left written information about their culture, including Mathematics and Astronomy, mainly in the form of inscriptions in stone sculptures—glyphs—now scattered throughout the ruins of the Maya empire in the Mesoamerica. Other written records, with very few exceptions, were destroyed by the Spaniards during the conquista.

The civilizations the Mathematics of which we are presenting here are far apart in both space and time (India 5000 BC till now—Mid and South America 1200 BC till about 1530 AD). It is quite amazing to observe that their Mathematics is essentially identical. This should make us think: Is Mathematics a universal idea for the entire humanity? The answer we propose is "yes"—there is one universal Mathematics in our world.

Acknowledgments should be made to Millersville University and West Chester University for granting us the sabbaticals that permitted us to begin the writing of this book. We are also indebted to Maya Exploration Center for offering many remarkable abroad mini-courses and workshops in Mexico, Guatemala, and Peru that we could attend. A special thanks to Drs. Ed Barnhart, Chris Powell, and Michael Grofe for leading the trips in which we could appreciate their many lectures that reassured our passion for these ancient cultures, and it made us rethink our view of mathematics and its relationship to culture.

We would like to stress that this book presents, in some cases, theories on how we believe the ancient conducted their computations to conceive their remarkable achievements.

Ximena Catepillán and Waclaw Szymanski

Preface by John B. Conway

It is a pleasure to introduce this book by Ximena Catepillán and Waclaw Szymanski. I find it an interesting and exciting approach to a semester long, terminal mathematics course for liberal arts students. Every university has such a course, some, like this book, well suited for the intended audience and some not. This book takes a special place in the available texts since not only does it accomplish the prime purpose of such courses, but it also exposes the student to different cultures and a historical perspective on mathematics.

In addition I find these authors highly qualified to undertake a book such as this. In addition to being excellent expositors and master mathematicians, I have always found them curious about and sensitive to other cultures. Their intellectual curiosity is intense, a requirement for an undertaking such as this. In fact it is difficult to imagine mathematicians with better backgrounds for such a project. In addition to their mathematical and intellectual virtues, they are committed to educating their readers. Professor Catepillán is a native of Chile and Professor Szymanski has spent considerable time in South and Central America, including a year teaching in Mexico City. So, both are familiar with and attuned to the Inca and Maya civilizations, whose mathematics forms the subject matter of most of the book. In addition, they are both scholarly. In one sense all mathematicians are scholars, but in the strictest meaning of the word most mathematicians are not. This is not a criticism of the practitioners of the field, but an acknowledgement that most mathematicians are more interested in the development or teaching of the subject than its origins. Professors Catepillán and Szymanski interests extend to the scholarship of the roots of the subject. Finally, this material has been tested in front of classes both at their universities and in classes of American students in Central America.

I learned things from reading this book. A striking example of this is that the concept of the zero number originated in India rather than the Arab world, as I had previously thought. More surprising, the Maya had the number zero, which they arrived at independently.

So I can easily recommend this book as a text for a course as well as a source of information and amusement for any mathematician.

Professor John B. Conway
Emeritus
Mathematics
George Washington University
Washington, DC

CHAPTER 1

The Language of Mathematics, Numbers, and Polynomials

1.1 The Language of Mathematics

Even though this book is mainly about Mathematics of ancient cultures, the modern language of Mathematics has to be used to make the book accessible to contemporary readers. Hence, some comments about the language of Mathematics are in order.

Let us begin with an example. Imagine yourself in the 4th or 5th grade vocabulary test. You are supposed to explain the meaning of the adjective "natural." Some possible explanations could be: "coming from nature," "original," "not artificial," "not human made," you may supply a few more. You are somehow familiar with the concept of a number. In Mathematics, the very first numbers you find are natural numbers. Out of our everyday knowledge of English, do we know what natural numbers are? Let us see: Are they "coming from nature," or "not artificial," or "not human made," or "original?" It all sounds wonderful, but do we really know what natural numbers are? The trouble with the language of Mathematics is that it looks like English, it sounds like English, but it is not English. One may say that Mathematics uses a foreign language disguised as English. In Mathematics, the words from ordinary language (English) are given entirely different meaning. This meaning of words is very precise, unlike in everyday language. Words are given meaning by definitions.

Now some definitions of mathematical words will be given to establish basic vocabulary needed to understand this book. For an easy reference, the definitions should be memorized, same way as the words of a foreign language have to be memorized to make understanding and speaking possible.

After the word "definition" the defined concept is **boldfaced**. The definition will be introduced in a natural way as the development of systems of numbers is presented.

1.2 Natural Numbers

Definition 1.
Natural numbers are: $1,\ 2,\ 3,\ 4, \ldots$(continue).

It is clear that 50 and 1000 are natural numbers, but -6 and $\frac{1}{2}$ are not. Natural numbers come from counting items. The concept of a number is abstract. The concept of a number of items is known to everyone. Suppose items are sticks. Then 2 represents the number of sticks pictured here: | |.

In the present day algebra, and **operation** in a set of numbers assigns to each two numbers in the set exactly one number in the same set.

Let us have a look at operations in the set of natural numbers. The basic operation is **addition**. Addition is the mathematical term for putting together, gathering together. This is the very foundation for counting items. Take two sticks | | and put them next to three sticks | | |. You get | |, | | | sticks. Put them together. You get | | | | | sticks. The number for | | | | | sticks is 5.

The notation we use now for addition (putting together) is $+$. Therefore, $| \; | + | \; | \; | = | \; | \; | \; | \; |$. Simply, the sign $+$ disappears. Using numbers we write $2+3=5$.

You may notice that polynomials are added the same way as sticks, just look at Example 1.6.3. Even though the concept of a polynomial came much later than the concept of a stick, the way we add in both cases is exactly the same.

If we add any two natural numbers, we get a natural number. Thus, addition is an operation in natural numbers. Indeed, if we put together a number of sticks (like | |) with another number of sticks (like | | |), we get a new number of sticks (| | | | |).

Definition 2.

a. Terms are numbers which are added.

b. The sum is the result of addition.

When we write $3+6=9$, the numbers 3 and 6 are terms and 9 is the sum.

Addition has two important properties; addition is:

- associative, which means $(a+b)+c=a+(b+c)$, and
- commutative, which means $a+b=b+a$, for all natural numbers $a,\ b,\ c$.

Let us see why these properties are natural. It all comes from putting sticks together.

Suppose we have 2 sticks | |, 3 sticks | | |, and 4 sticks | | | |. Then $(2+3)+4=$ (| | + | | |) + (| | | |) = | | | | | + | | | | = | | | | | | | | | = 9. Just drop the + sign. On the other hand, $2+(3+4)=$ | | + (| | | + | | | |) = | | + (| | | | | | |) = 9. This is why addition is associative.

Now, $2+3=$ | | + | | | = | | | | | = | | | + | | = $3+2$. This makes it obvious that addition is commutative.

Usually, after addition, subtraction is discussed. Let us have a closer look at subtraction. Is subtraction an operation in natural numbers? Is it true that if we subtract any two natural numbers, we will get a natural number? What about $3-9$? Is there a natural number assigned to $3-9$? The answer is, clearly, NO. Thus, subtraction is not an operation in natural numbers. In natural numbers, we can always subtract a smaller number from a larger one, but this is not enough to treat subtraction as an operation. For subtraction as an operation, a larger set of numbers is needed.

The second operation in natural numbers is **multiplication**. If two natural numbers are multiplied, the result is a natural number. By definition, we multiply two natural numbers $m,\ n$ by adding n to itself m times. For example, if $m=3$, $n=6$ then we multiply them by adding three copies of $6: 6+6+6$.

Definition 3.

a. Factors are numbers which are multiplied.

b. The product is the result of multiplication.

The sign for multiplication is: "×" or "·" between two numbers, (3×6 or $3\cdot 6$) and nothing between two letters or a number and a letter ($xy, 3x$). The sign "·" can always be used: $xy=x\cdot y,\ 3x=3\cdot x$. In the last case, do not use "×" as it resembles the letter x, and this may cause mistakes.

When we write $3\times 6=18$ or $3\cdot 6=18$, the numbers 3 and 6 are factors and 18 is the product.

The definition of multiplication determines the order of operations. Suppose we want to find $3\cdot 6+9$. We get $3\cdot 6=6+6+6$ and $3\cdot 6+9=6+6+6+9=27$. Hence, **the order of operations is: multiplication before addition.**

Multiplication has two properties similar to properties of addition. Multiplication is:

- associative, which means $(ab)c=a(bc)$, and
- commutative, which means $ab=ba$.

Addition and multiplication are connected by the distributive property: $a(b+c)=ab+ac$,

which means multiplication distributes over addition.

These properties are easily justified using definition of multiplication and properties of addition.

Now let us look at division. It is clear that there is no natural number assigned to $3 \div 6$. Thus, the division as an operation does not exist in natural numbers.

There is, however, an important theorem, which will be used several times later in this book. The theorem is called The Division Algorithm. It justifies the procedure of "division with a reminder."

Let us begin with two examples:

1. Suppose a mother has 15 apples and wants to distribute evenly to three children. It is obvious that each child will get $15 \div 3 = 5$ apples. This means $15 = 3 \times 5 + 0$.
2. Now suppose there are 16 apples to be distributed to three children. Then each child gets 5 apples and 1 apple is left (no apple is cut). This means $16 = 3 \times 5 + 1$.

These are examples of division with remainder. Remainder is what is left. In example 1 the remainder is 0, and in example 2 the remainder is 1.

The division algorithm assures that such procedure is always possible. It requires a proof (not given here) which is based on properties of natural numbers.

Division Algorithm.
Suppose a, b are natural numbers. Then unique natural numbers q, r can be found so that:

a. $a = bq + r$, and
b. $r < b$.

Numbers q or r can also be 0, but not both.

Usually a is called the dividend, b is the divisor, q is the quotient, and r is the remainder.

Therefore, conditions (a) and (b) of Division Algorithm can be written as:

a. dividend = divisor × quotient + remainder,
b. remainder < divisor.

In both examples 1 and 2, $a = 15$, $b = 3$, $q = 5$.

1.3 Integer Numbers

Definition 4.
Integer numbers (integers) are: $\ldots, -3, -2, -1, 0, 1, 2, 3, \ldots$

The dots mean keep going like that to the left and to the right. In integers, there is addition and multiplication with all properties listed for natural numbers. The most important concept for integers, totally absent in natural numbers, is the concept of the opposite.

Definition 5.
A number b is **the opposite** of a number a if $a + b = 0$.

The opposite of 5 is -5, the opposite of -3 is 3, the opposite of 0 is 0.

The symbol for the opposite of a number a is $-a$. Notice that $-a$ does not have to be negative. If $a = 5$, then $-a = -5 < 0$, but if $a = -3$ then $-a = 3 > 0$. In general, if $a > 0$ then $-a < 0$, and if $a < 0$ then $-a > 0$.

The concept of the opposite brings the operation of subtraction.

Definition 6.
To subtract means to add the opposite.

This definition means: to subtract number b from number a, we add the opposite $-b$ of b to a.

$a-b=a+(-b)$.

Since each integer has an opposite, we have subtraction as an operation. The reason to introduce integers is to make subtraction possible.

For the same reason as in natural numbers, there is no operation of division in integers.

Now we introduce a set of numbers, in which there is also the operation of division.

1.4 Rational Numbers

Definition 7.
A **rational number** is a ratio of integers.

Ratios are often expressed as fractions, thus a rational number is a fraction $\frac{a}{b}$ where a, b are integers, $b \neq 0$. Incidentally, the name "rational" is derived from "ratio."

Integer numbers are rational, because, for example: $2=\frac{2}{1}$, $5=\frac{50}{10}$ are ratios of integers. Numbers $\frac{1}{2}, \frac{3}{5}, \frac{-8}{3}$ are rational, but not integer.

In rational numbers, there are the operations of addition, subtraction (usually the common denominator is needed), and multiplication. The most important concept in rational numbers is the concept of the reciprocal.

Definition 8.
A number b is **the reciprocal** of a number a if $ab=1$.

The reciprocal of 2 is $\frac{1}{2}$, the reciprocal of $-\frac{1}{3}$ is -3, reciprocal of $\frac{3}{5}$ is $\frac{5}{3}$, the reciprocal of 1 is 1. The symbol for the reciprocal of a number a is $\frac{1}{a}$. The concept of reciprocal brings in the operation of division.

Definition 9.
To divide means to multiply by the reciprocal.

This definition means: to divide a number a by a number b, we multiply a by the reciprocal $\frac{1}{b}$ of b.

$a \div b = a \cdot \frac{1}{b}$.

If we want to divide by a number, this number has to have a reciprocal. Looking at the definition of rational numbers when we pick one (like $-\frac{2}{7}$), we can immediately give its reciprocal $(-\frac{7}{2})$. So it seems that we can divide by any rational number. Is it really true? Weren't we told from grade school that division by 0 is forbidden? But why? In Mathematics, everything must have an explanation, if we could divide by 0 then 0 has to have a reciprocal. But then, which number is the reciprocal of 0? Perhaps there is not any. Yes. We formulate this as a theorem—a statement which requires justification (called a proof). We recommend that you learn and remember the following proof.

Theorem 1.4.1.
0 has no reciprocal.

Proof.
Suppose 0 has a reciprocal and see what happens. Let b be the reciprocal of 0. Let us ask how much is $0 \cdot b$?

This question has two answers:

Answer 1: $0 \cdot b = 0$, because $0 \cdot$ (any number) is 0.

Answer 2: $0 \cdot b = 1$, because b is the reciprocal of 0—just look carefully at Definition 8 for $a = 0$.

From Answer 1 and Answer 2, the only conclusion is that $0 = 1$. Thus, we proved that if 0 has a reciprocal then $0 = 1$. We know that $0 \neq 1$. Hence, 0 has no reciprocal. End of proof. ∎

Theorem 1.4.1 shows that division by 0 is impossible, because 0 has no reciprocal. It also explains why the denominator of a fraction cannot be 0. All other rational numbers have reciprocals. Thus, in rational numbers, there is operation of division, except by 0. In fact, rational numbers are introduced to make division possible (except by 0).

In rational numbers, we have all four arithmetic operations: addition, subtraction, multiplication, and division (except by 0). Do we need any more numbers?

Any measurements in the physical world (length, area, temperature, etc.) are always expressed in terms of rational numbers. In Mathematics—even just elementary geometry—we find numbers which are not rational, in very simple situations.

1.5 Real Numbers

Let us recall the best known theorem in geometry.

Pythagorean Theorem.
In a right triangle, the square of the longest side equals the sum of squares of the remaining sides.

Warning. Pythagorean Theorem is NOT $a^2 + b^2 = c^2$ without explaining what are a, b, c.

Now consider the square with length of side 1 and diagonal d. You see here two identical (congruent) right triangles with the longest side (hypothenuse) d. Pythagorean theorem here says: $d^2 = 1^2 + 1^2$,which is: $d^2 = 2$. Ancient Greeks, especially Pythagoras, knew that if $d^2 = 2$ then d is not a rational number.

Definition 10.
A number b is the **square root** of a number $a \geq 0$ if $b^2 = a$ and $b \geq 0$.

The notation for the square root of a is $\sqrt{a}$.

Using this concept, the last statement concerning the diagonal of the square above can be written as: $\sqrt{2}$ is not rational.

Definition 11.
A natural number is **prime** if its only natural factors are 1 and itself.

It is proved that the square root of any prime number is not rational.

Here are some prime numbers: $2, 3, 5, 7, 11, 13, 17, 19, \ldots$.

A Greek philosopher-mathematician Euclid proved that there are infinitely many prime numbers. Since the square root of each of them is not rational, we have infinitely many numbers that are not rational.

One more important number coming from geometry is not rational either, it is π.

Definition 12.
π is the ratio of the circumference of a circle to its diameter.

Warning. π is not 3.14, because 3.14 is rational and π is not rational.

The real numbers are introduced to contain rational numbers, square roots of prime numbers, π, and a lot more. It is impossible to give the definition of real numbers without much more knowledge of Mathematics.

One of the most important reasons real numbers are introduced is that they can be identified with points on a line in such a way that the line is solid—has no holes; such line with that identification is called real line. A real line is a basic tool of Mathematics. (Szymanski 1983)

1.6 Powers, exponents, polynomials

Suppose a is a real number, n is a natural number.

Definition 13.
The $n-th$ **power** of a is the product of a by itself n times.

The symbol for the $n-th$ power of a is a^n. Here a is called the base and n is called the exponent. Using this symbol, we write the above definition as follows:

$a^n = a \cdot a \cdot a \ldots a$ (or a repeated n times).
n times

For example: $2^4 = 2 \cdot 2 \cdot 2 \cdot 2 = 16,\ 3^2 = 3 \cdot 3 = 9$. If a number a has no exponent, it means that the exponent is 1, that is $a = a^1$.

Definition 14.
$a^0 \equiv 1$ if $a \neq 0$.

Exponents have several properties. Here we need only one. Suppose m, n are natural numbers or 0. Then,

Property 1.6.1
$a^m a^n = a^{m+n}$

This property says that to multiply two powers with the same base we add the exponents.

Let us see that this property is true for $m = 2,\ n = 3$.

$a^m a^n = a^2 a^3 = (a \cdot a) \cdot (a \cdot a \cdot a) = a \cdot a \cdot a \cdot a \cdot a = a^5 = a^{2+3} = a^{m+n}$.

Now we discuss polynomials. In algebra, letters are used to represent numbers. We perform operations on letters, like $x + y$, or x^2. Such expressions are called algebraic expressions. They can be evaluated for particular values of letters. For example, if $x = 2,\ y = 6$ then $x + y = 2 + 6 = 8$.

Definition 15.
- **a.** A **monomial** is a product of letters and numbers.
- **b.** A **polynomial** is a sum or difference of monomials.

Recall Definition 3.b: the product is the result of multiplication. Therefore, $2 = 2 \cdot x^0, x^2 = x \cdot x, 3x, 4x^2y$ are examples of monomials.

The product of monomials is a monomial.

Example 1.6.2
- **a.** $x \cdot (3x) = x^1 \cdot 3x^1 = 3x^{1+1} = 3x^2$
- **b.** $(5x)^2 = (5x) \cdot (5x) = 5x^1 \cdot 5x^1 = 5 \cdot 5x^{1+1} = 25x^2$
- **b.** $(3xy) \cdot (4x^2y^3) = 3x^1y^1 \cdot 4x^2y^3 = 3 \cdot 4x^{1+2}y^{1+3} = 12x^3y^4$

We use here property 1.6.1 and commutativity of multiplication.

In polynomials, we have operations of addition, subtraction, and multiplication. We will explain these operations on examples.

Addition.

Example 1.6.3

$(3x^2+2x-5)+(4x-6)=3x^2+2x-5+4x-6=3x^2+2x+4x-5-6=3x^2+6x-11.$

To add polynomials, we drop the parentheses and get the sum which is a polynomial. Since the addition is commutative, terms can be rearranged and terms with equal powers of x can be added. It follows from the distributive property (in 1.2) that $2x+4x=(2+4)x=6x$.

Subtraction.

The difference is the result of subtraction. According to Definition 6, to subtract means to add the opposite. Let us find the opposite of $x-y$. Apply Definition 5 with $a=x-y$. We are looking for $b=(\quad)$ such that $a+b=0$, that is, $(x-y)+(\quad)=0$.

It is clear that $b=-x+y$, because $(x-y)+(-x+y)=x-y-x+y=0$ (addition is commutative). This brings about the rule:

Rule 1.6.4

To get the opposite of a polynomial, change sign at each term of the polynomial.

Example 1.6.5

$(3x^2+2x-5)-(4x-6)=(3x^2+2x-5)+(-4x+6)=3x^2+2x-5-4x+6=3x^2-2x+1.$

Multiplication.

Multiplication of polynomials is based on the distributive property stated in 1.2.: $a(b+c)=ab+ac$. Here is the distributive property in an example.

Example 1.6.6

$2x^3(3x^4+4)=2x^3\cdot 3x^4+2x^3\cdot 4=6x^7+8x^3.$

In general, distributive property has to be used several times.

Example 1.6.7

$(2x^3+x)(3x^4+4)=(2x^3+x)\cdot 3x^4+(2x^3+x)\cdot 4=2x^3\cdot 3x^4+x\cdot 3x^4+2x^3\cdot 4+x\cdot 4=6x^7+3x^5+8x^3+4x.$

The distributive property was used here three times. The expression $2x^3\cdot 3x^4+x\cdot 3x^4+2x^3\cdot 4+x\cdot 4$ brings about the following general rule:

Rule 1.6.8

To multiply two polynomials, multiply each term of the first polynomial by each term of the second polynomial.

Example 1.6.9

$(a+b)^2=(a+b)(a+b)=a^2+ab+ba+b^2=a^2+2ab+b^2.$

Since multiplication is commutative: $ab=ba$.

Example 1.6.10

Here, we will use Example 1.6.9.

$(a+b)^3=(a+b)(a+b)(a+b)=[(a+b)(a+b)](a+b)=(a^2+2ab+b^2)(a+b)=a^3+a^2b+2a^2b+2ab^2+b^2a+b^3=a^3+3a^2b+3ab^2+b^3.$

Example 1.6.11
$(2x+5)(3x+4) = 2x\cdot 3x + 2x\cdot 4 + 5\cdot 3x + 5\cdot 4 = 6x^2 + 8x + 15x + 20 = 6x^2 + (8+15)x + 20 = 6x^2 + 23x + 20.$

Here, like in Example 1.6.3, we used distributive property $8x + 15x = (8+15)x$ to get the number multiplying x.

Example 1.6.12
$(x^2 - 2x + 3)(4x^2 + 5x - 6) = x^2\cdot 4x^2 + x^2\cdot 5x + x^2\cdot(-6) - 2x\cdot 4x^2 - 2x\cdot 5x - 2x\cdot(-6) + 3\cdot 4x^2 + 3\cdot 5x + 3\cdot(-6) =$

$4x^4 + 5x^3 + (-6)x^2 - 8x^3 - 10x^2 + 12x + 12x^2 + 15x - 18 = 4x^4 + 5x^3 - 8x^3 + (-6)x^2 - 10x^2 + 12x^2 + 12x + 15x - 18 =$

$4x^4 + (5-8)x^3 + (-6-10+12)x^2 + (12+15)x - 18 = 4x^4 + (-3)x^3 + (-4)x^2 + 27x - 18.$

Here we have used the distributive property: $5x^3 - 8x^3 = (5-8)x^3$, $(-6)x^2 - 10x^2 + 12x^2 = (-6-10+12)x^2$, and $12x + 15x = (12+15)x$.

1.7 Roots of Polynomials. Degree 1

Here and in 1.8, we will discuss polynomials of one variable x. According to Definition 15.a, a monomial is a product of letters and numbers. Since there is only one letter x, the product of x by itself a certain number of times is a power of x, like $x^3 = x\cdot x\cdot x$ —just look at Definition 13. Such power may be multiplied by a number, so monomials in one variable are like $2 = 2x^0$, $3x$, $5x^2$, $4x^3$. Definition 15.b says that a polynomial is a sum or difference of monomials. Hence, polynomials here are expressions like: 2, $3x+2$, $5x^2 - 3x + 2$.

Each monomial is a polynomial, because you may always add 0, for example $3x = 3x + 0$. The **degree** of a polynomial is the highest power of x. Hence, 2 is of degree 0, $3x+2$ of degree 1, $5x^2 - 3x + 2$ of degree 2.

Definition 16.
A number is a **root** of a polynomial if when the variable x is replaced by that number, we get 0.

For example, the root of the polynomial $x+2$ is -2, because $-2+2=0$.

Finding roots of polynomials is the same as solving equations.

When we move an expression from one side of an equation to the other, the expression changes sign. The reason is the following property of equality: if $a=b$ then $a+c=b+c$. Hence, if $4x-2=0$, then $4x=2$, because 2 was added to both sides. Similarly, if $a=b$, then $ac=bc$ for $c\neq 0$. Hence, if $4x=2$, then $4x\cdot\frac{1}{4} = 2\cdot\frac{1}{4}$ thus $x=\frac{2}{4}=\frac{1}{2}$. Check: substitute $\frac{1}{2}$ for x to $4x-2$. You get $4\cdot\frac{1}{2} - 2 = 2 - 2 = 0$.

This is the way we find roots of polynomials of degree 1.

Example 1.7.1
Solve

$$\begin{aligned} 5x + 2 &= 3x - 1 \\ 5x - 3x &= -1 - 2 \\ 2x &= -3 \\ x &= -\frac{3}{2} \end{aligned}$$

Example 1.7.2
Solve $5(x+2) = 3(x-1)$

This time use distributive property on both sides to get

$5x + 10 = 3x - 3$

Then follow Example 1.7.1

$$\begin{aligned} 5x - 3x &= -3 - 10 \\ 2x &= -13 \\ x &= -\frac{13}{2} \end{aligned}$$

1.8 Roots of Polynomials. Degree 2 - Quadratic

The simplest polynomial of degree 2 is x^2.

Suppose the equation to solve is $x^2 = 4$. It is clear that $x = 2$ or $x = -2$ are solutions. In general,

Example 1.8.1
Let d be a real number. The solutions of the equation $x^2 = d$ are:

a. if $d > 0$, then $x = \sqrt{d}$ or $x = -\sqrt{d}$; this follows from Definition 10,
b. if $d = 0$, then $x = 0$; this is clear,
c. if $d < 0$, then there is no solution, because $x^2 \geq 0$ for each real number x.

The general quadratic polynomial is $ax^2 + bx + c$ with $a \neq 0$. The most important number associated with this polynomial is **the discriminant** defined by $\Delta = b^2 - 4ac$. The symbol Δ is the Greek capital letter delta.

Example 1.8.2

a. $2x^2 + 5x - 10$. Here $a = 2,\ b = 5,\ c = -10$. Hence, $\Delta = b^2 - 4ac = 5^2 - 4 \cdot 2 \cdot (-10) = 25 + 80 = 105$.
b. $x^2 - 4x + 4$. Here $a = 1,\ b = -4,\ c = 4$. Hence, $\Delta = b^2 - 4ac = (-4)^2 - 4 \cdot 1 \cdot 4 = 16 - 16 = 0$.
c. $-x^2 + 2x - 5$. Here $a = -1,\ b = 2,\ c = -5$. Hence, $\Delta = b^2 - 4ac = 2^2 - 4 \cdot (-1) \cdot (-5) = 4 - 20 = -16$.

The way the roots of a quadratic polynomial are found in all textbooks come from the Greeks. The justification requires a process called "*completing the square*."

Here is a different approach, much easier and more natural than "*completing the square*." It comes from ancient Hindus and for that reason we call it Vedic approach. This name and historical background are explained at the beginning of the next chapter "Vedic Mathematics." (Jagadguru 2010, Datta 2004)

Let us call our polynomial y, that is, $y = ax^2 + bx + c,\ a \neq 0$. Define a new polynomial $y' = 2ax + b$. (If you know some elementary calculus, y' is the derivative of y, hence the notation.) Now let us make a simple calculation—square y'.

$(y')^2 = (2ax + b)^2 =$ (use Example 1.6.9) $= 4a^2x^2 + 4abx + b^2 =$ (add and subtract $4ac$) $= 4a^2x^2 + 4abx + 4ac + b^2 - 4ac =$ (group terms) $= (4a^2x^2 + 4abx + 4ac) + (b^2 - 4ac) =$ (use distributive property in the first group—take out $4a$) $= 4a(ax^2 + bx + c) + (b^2 - 4ac) = 4ay + \Delta$.

This establishes the following equation, which we call Vedic Equation for Quadratic Polynomials (VEQP).

(VEQP) $\qquad (y')^2 = 4ay + \Delta$

From this equation, we get roots of y without effort. Definition 16 says that a root is a number x for which $y = 0$. Hence, substitute $y = 0$ to (VEQP). We get $(y')^2 = \Delta$, which means $(2ax + b)^2 = \Delta$. Now we use Example 1.8.1 replacing x there by $2ax + b$ and d by Δ. This way we get:

Theorem 1.8.3.

a. If $\Delta > 0$, then $2ax + b = \sqrt{\Delta}$ or $2ax + b = -\sqrt{\Delta}$.
b. If $\Delta = 0$, then $2ax + b = 0$.
c. If $\Delta < 0$, then there is no solution, which means, y has no root.

Notice that what is written before this theorem is a complete proof of it.

Let us see how this theorem works in practice. In the next example, we use what we found in Example 1.8.2.

Example 1.8.4

a. $y = 2x^2 + 5x - 10,\ \Delta = 105,\ y' = 2ax + b = 4x + 5$
Hence Theorem 1.8.3 (a) has to be used. Thus

$$2ax + b = \sqrt{\Delta} \quad \text{or} \quad 2ax + b = -\sqrt{\Delta}$$
$$4x + 5 = \sqrt{105} \quad \text{or} \quad 4x + 5 = -\sqrt{105}$$

To solve these equations just do what we did in Example 1.7.1.

$$4x = -5 + \sqrt{105} \qquad 4x = -5 - \sqrt{105}$$
$$x = \tfrac{-5+\sqrt{105}}{4} \qquad x = \tfrac{-5-\sqrt{105}}{4}$$

These two numbers are the roots of y.

b. $y = x^2 - 4x + 4,\ \Delta = 0,\ y' = 2ax + b = 2x - 4.$
Thus Theorem 1.8.3 (b) will be used:

$$2ax + b = 0$$
$$2x - 4 = 0$$
$$2x = 4$$
$$x = 2$$

Hence, $x = 2$ is a root of y. This can be checked by substituting $x = 2$ to y, $2^2 - 4 \cdot 2 + 4 = 4 - 8 + 4 = 0$.

c. $y = -x^2 + 2x - 5,\ \Delta = -16.$
According to Theorem 1.8.3 (c), y has no root.

Let us look at more examples.

Example 1.8.5
$y = 3x^2 - 2x - 1$

Here $a = 3,\ b = -2,\ c = -1$. Hence, $\Delta = b^2 - 4ac = (-2)^2 - 4 \cdot 3 \cdot (-1) = 4 + 12 = 16$ and $y' = 2ax + b = 6x - 2$.

We use Theorem 1.8.3 (a). Notice that $\sqrt{\Delta} = \sqrt{16} = 4$

$$2ax + b = \sqrt{\Delta} \qquad 2ax + b = -\sqrt{\Delta}$$
$$6x - 2 = 4 \qquad 6x - 2 = -4$$
$$6x = 6 \qquad 6x = -2$$

Roots: $x = 1$ and $x = -\frac{2}{6} = -\frac{1}{3}$.

Example 1.8.6
Sometimes the question is: solve the equation $-2x^2 - 1 = -4x$.

In such case we move all terms to one side leaving 0 on the other

$-2x^2 - 1 + 4x = 0.$

We have to make the polynomial on the left look like $ax^2 + bx + c$, therefore, we rearrange terms:

$-2x^2 + 4x - 1 = 0.$

Now the question to solve the equation is changed to the question to finding roots of $y = -2x^2 + 4x - 1$.

Here $a=-2,\ b=4,\ c=-1$. Hence, $\Delta=b^2-4ac=4^2-4\cdot(-2)\cdot(-1)=16-8=8$ and $y'=2ax+b=-4x+4$.

Now Theorem 1.8.3 (a) again gives

$$\begin{aligned} 2ax+b &= \sqrt{\Delta} & \qquad 2ax+b &= -\sqrt{\Delta} \\ -4x+4 &= \sqrt{8} & -4x+4 &= -\sqrt{8} \\ -4x &= -4+\sqrt{8} & -4x &= -4-\sqrt{8} \\ \text{Roots:}\quad x &= \tfrac{-4+\sqrt{8}}{-4} & x &= \tfrac{-4-\sqrt{8}}{-4} \end{aligned}$$

Example 1.8.7
$y=-9x^2+6x-1$

Here $a=-9,\ b=6,\ c=-1$. Hence, $\Delta=b^2-4ac=6^2-4\cdot(-9)\cdot(-1)=36-36=0,\ y'=2ax+b=-18x+6$.

Now Theorem 1.8.3 (b) gives

$$\begin{aligned} 2ax+b &= 0 \\ -18x+6 &= 0 \\ -18x &= -6 \\ \text{Root:}\quad x &= \tfrac{-6}{-18} = \tfrac{1}{3} \end{aligned}$$

Example 1.8.8
Solve the equation $x^2-x=-1$.

Add 1 to both sides to get $x^2-x+1=0$. Hence, we look for roots of $y=x^2-x+1$. Here $a=1,\ b=-1,\ c=1$. Hence, $\Delta=b^2-4ac=(-1)^2-4\cdot1\cdot1=1-4=-3$. Thus Theorem 1.8.3 (c) tells that there is no root, so the equation has no solution.

EXERCISES

1. Use sticks to give three more examples that addition is associative and commutative.
2. Check properties of multiplication and the distributive property for $a=2, b=3$, and $c=4$.
3. Apply Division Algorithm to find the quotient and the remainder for:

	Dividend	Divisor
a.	127	25
b.	1,024	64
c.	2,765	26
d.	3,812	31
e.	4,015	33

4. Perform the operations.

a. $3-4\cdot5$ **b.** $(3-4)\cdot5$ **c.** $2+3\cdot5-7$
d. $(2+3)\cdot5-7$ **e.** $2-4\cdot5+3$ **f.** $(2-4)\cdot(5+3)$

5. List all prime numbers up to 100.
6. Perform all operations.

a. $x^2\cdot(2x)$ **b.** $(6x)^2$ **c.** $(2x^2y)\cdot(3xy^4)$
d. $(3x^2+4x-1)+(3x-2)$ **e.** $(3x^2+4x-1)-(3x-2)$ **f.** $(3x^2+4x-1)\cdot(3x-2)$

g. $(2x^3 - 5x^2 + 1) + (x^3 - 2)$ **h.** $(2x^3 - 5x^2 + 1) - (x^3 - 2)$ **i.** $(2x^3 - 5x^2 + 1) \cdot (x^3 - 2)$

j. $(x - y)^2$ **k.** $(x - y)^3$

7. Find roots of the polynomials and check.

a. $3x + 6$ **b.** $3x - 8$ **c.** $5x - 1$

8. Solve the equations and check.

a. $2x + 4 = 10$ **b.** $3x - 7 = -5$ **c.** $x - 1 = 3x + 2$

d. $4x - 3 = 2x + 1$ **e.** $2(x + 4) = 10$ **f.** $3(x - 7) = -5$

g. $x - 1 = 3(x + 2)$ **h.** $4(x - 3) = 2(x + 1)$

9. Solve the equations.

a. $x^2 = 9$ **b.** $x^2 = -9$ **c.** $x^2 = 5$

d. $x^2 = -5$ **e.** $(3x + 1)^2 = 0$ **f.** $(3x + 1)^2 = 2$

g. $(3x + 1)^2 = -2$ **h.** $(2x - 5)^2 = 0$ **i.** $(2x + 5)^2 = 4$

j. $(2x - 5)^2 = -4$

10. For the following polynomials, identify a, b, c, compute Δ, y', and use the Vedic method (Theorem 1.8.3) to find roots of the polynomials.

a. $y = x^2 - x - 1$ **b.** $y = x^2 - x + 1$ **c.** $y = 3x^2 - 2x + 1$

d. $y = 3x^2 - 2x - 1$ **e.** $y = x^2 + 4x + 4$ **f.** $y = x^2 + 4x - 4$

g. $y = 9x^2 - 6x + 1$ **h.** $y = 9x^2 - 6x - 1$ **i.** $y = 2x^2 - 4x - 1$

j. $y = 2x^2 - 4x - 1$ **k.** $y = 2x^2 - 4x + 1$ **l.** $y = -x^2 - x + 1$

m. $y = -x^2 - x - 1$ **n.** $y = -3x^2 + x + 2$ **o.** $y = -3x^2 + x - 2$

11. For each equation, find the polynomial $y = ax^2 + bx + c$, like in Example 1.8.6 or 1.8.8. Then follow the directions of Exercise 10 above.

a. $-x^2 + 2 = x$ **b.** $y = 3x^2 + x = 1$ **c.** $3x^2 + x = -1$

d. $6x + 9 = -x^2$ **e.** $6x + 9 = x^2$ **f.** $2x^2 + x = 6$

g. $2x^2 + x = -6$

REFERENCES

Jagadguru Swami Sri Bharati Krsna Tirthaji Maharaja. 2010. *Vedic Mathematics.* 13th ed. Delhi, India: Motilal Banarsidass Publishers.

Datta, Bibhutibhushan and Avadesh Narayan Singh. 2004. *History of Hindu Mathematics.* Vol. II. Delhi, India: Bharatiya Kala Prakashan.

Szymanski, Waclaw. 1983. Introduction to Mathematics part I - Elements of Set Theory and Basic Concepts of Algebra (Polish). Krakow, Poland: Academy of Agriculture.

CHAPTER 2

Vedic Mathematics

उत्पादकं यत्प्रवदन्ति बुद्धेरधिष्ठितं सत्पुरुषेण सांख्याः ।
व्यक्तस्य कृत्स्नस्य तदेकबीजम् अव्यक्तमीशं गणितं च वन्दे ।।

Devanagari text of the First Verse of Bījaganita of Bhāskarācārya (c. 1150)

I salute that Īśa (the ruling power, Brahman), which Sānkhyas (those who have realised the Self) declare to be the producer of buddhi (true knowledge of reality), which arises in a distinguished person. It is the sole bīja (seed or the cause) of all that is manifest.
I salute that avyakta-ganita (the mathematics with unmanifest or indeterminate quantities), which the Sānkhyas (those who are proficient in numbers) declare to be the producer of buddhi (mathematical knowledge), which arises in a distinguished person. It is sole bīja (seed or the cause) of all vyakta-ganita (the mathematics with manifest quantities, such as arithmetic, geometry etc.). (First Verse of Bījaganita of Bhāskarācārya (c. 1150) with two meanings.)

2.1 Historical Background

Indus

Indus is the main river in the present day Pakistan. Throughout the history the territory Pakistan was a part of India. Pakistan as a country was created only in 1947—the year India gained independence from the British. Indus is 1990 miles long, which makes it one of the longest rivers in Asia. Indus originates in the glaciers of the Himalayas in Tibet near the famous lake Mansarovar. It flows through the entire of Pakistan and ends in the Arabian Sea near the port of Karachi. Sanskrit (see remarks on Sanskrit below) name for Indus is Sindhu.

The Valley of Indus is the cradle of Hindu civilization. This is the only civilization in the world, which exists continuously, practically unchanged, for over 5000 years (Figure 2.1).

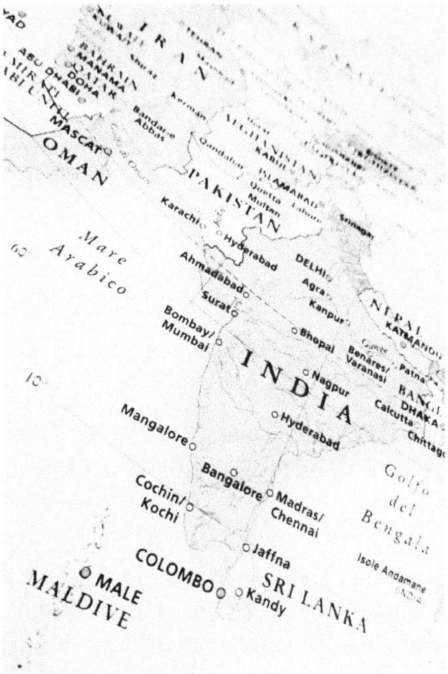

© TonelloPhotography/Shutterstock.com

Figure 2.1: Map of India

Harappan civilization (c. 3000 BC–1900 BC)

Archaeologists discovered ruins of two large cities in the Valley of Indus: Harappa and Mohenjo-Daro. There are ruins of smaller settlements as well in that region. From the name of the city of Harappa comes the name Harappan civilization. The two large, impressive cities were precisely planned and laid out. Just from this

© noppharat/Shutterstock.com

Figure 2.2: Manuscript in Sanskrit

example historians draw the conclusion that Harappan civilization must have a knowledge of Mathematics, architecture, and engineering. Harappan civilization started to decline about 2000 BC.

Aryans

The topic of the so-called Aryan invasion is highly controversial. Some European historians specializing in the history of the East formulated in the nineteenth century the Aryan Invasion Theory. According to this theory, around 1900 BC the tribe of the Aryans invaded the Harappan civilization and caused its decline. This theory is now treated by many historians in India and in the West merely like an unproven hypothesis. The very existence of Aryans is being questioned. The serious archaeological excavations in Harappa and Mohenjo-Daro started only in the twentieth century and it is claimed that up to now no evidence of the Aryans was found. (Malhotra, 2011)

Vedas

The word *Veda* is a Sanskrit word which means knowledge. The word *Vedic* is the English adjective of Veda.

The *Vedas* (English plural of Veda) are the sacred texts of the Hindus. There are four principal Vedas: Rig Veda, Sama Veda, Yajur Veda, and Atharva Veda.

The Vedas are the foundation of the Hindu civilization. They contain the knowledge of every aspect of the human life including descriptions of rituals, chants, hymns, and very profound philosophical system contained in the part called Vedanta. The Vedas have no authors, they do not come from humans. Some believe that God communicated the knowledge to the ancient sages. Nobody knows with certainty how and when the Vedas really originated.

This ancient knowledge was first passed from the teacher to the student only orally. The teacher chanted a passage and the student repeated. The student had to memorize the text chanted by the teacher whether he understood it or not. Understanding of the text came later—at a higher level of learning. Later the system of writing was invented using the set of symbols—script—called Devanagari. The Vedas were recorded using this system. Writing was preserved on palm leaves. Since the time immemorial, the Vedas were never modified. Today these sacred texts are exactly the same as in the beginning. (See (Figure 2.2)

© Nila Newsom/Shutterstock.com

Figure 2.3: Sanskrit text found on a golden plate at a temple in Nepal

Sanskrit

The Sanskrit word *Sanskrit* means "well made." Sanskrit is the oldest and only language of the world preserved intact until the present time. First it was only spoken; when the writing was invented, Sanskrit used the Devanagari script. It has mathematically precise grammar. Almost all European languages have their roots in Sanskrit. This is why they are called Indo-European languages. The language of the Vedas is Sanskrit (Figure 2.3).

2.2 Vedic Mathematics—Introduction

The Sanskrit word for Mathematics is *ganita*. Literally it means "the science of calculation." This word appears numerous times in Vedic literature. The Yajurveda (c. 600 BC)—one of the four principal Vedas—gives Mathematics the highest place among all sciences:

As the crest on the heads of peacocks, as the gems on the hoods of cobras, so is ganita at the top of the sciences known as the vedanga.

Vedic Mathematics is the name given to the ancient system of Hindu Mathematics. It was rediscovered from the Vedas between 1911 and 1918 by Sri Bharati Krsna Tirthaji Maharaja. According to his opinion, all of Mathematics is based on sixteen sutras. The Sanskrit word sutra means aphorism, short, terse statement. In Mathematics, it can be interpreted as "word formula." Sutras have to be interpreted and explained.

The sixteen sutras on which Mathematics is based can be found in Table 2.1.

These formulae describe the way the mind naturally works. Therefore, they are very helpful in directing students to the appropriate method of solution of a problem.

The sutras are quite universal. They can be applied in various areas of Mathematics, not only in arithmetic. They also can be used in a broader sense beyond Mathematics, for example, in physics or astronomy.

Table 2.1: The sixteen sutras

Sutra	Sanskrit	Name
1	Ekadhikena	By one more than the one before
2	Nikhilam Navatashcharamam	All from 9 and the last from 10
3	Urdhva-tiragbhyam	Vertically and cross-wise
4	Paravartya Yojayet	Transpose and apply
5	Shunyam Samyasamuchchaye	When the sum is the same that sum is zero
6	Anurupye shunyamanyat	If one is in ratio the other is zero
7	Sankalana vyavakalanabhayam	By addition and by subtraction
8	Purnapurnanabhyam	By the completion or noncompletion
9	Calana-kalana	Calculating variation
10	Yavadunam	By the deficiency
11	Vyashtisamashtihi	Specific and general
12	Seshayankena Charamena	The remainders by the last digit
13	Sopantyadvayamamantyam	The ultimate and twice the penultimate
14	Ekanyunena Purvena	By one less than the one before
15	Gunitasamuchchayaha	The product of the sum
16	Gunakasamuchchayaha	All the multipliers

Here is an easy example. Suppose, we want to add 156 and 98. Instead of reaching for a calculator, think a moment. Let us add 156 and 100 to get 256. Then subtract 2 to get the final result 254. Hence, $156 + 98 = 254$. This is a natural mental method. It uses the observation that 98 is deficient from 100 by 2. This example illustrates sutra 10—By the deficiency.

Keep in mind that computers and calculators are not even 100 years old. Yet, people were able to perform sometimes very complicated arithmetic operations simply in their head—mentally—thousands of years ago.

On p. 231 of Vedic Metaphysics, Sri Bharati Krsna Tirthaji Maharaja, 2010, states that he found all sixteen sutras in Shtapataya Veda in connection with astronomy. It is quite feasible that this is not a published source. There are ancient texts commonly accepted as Vedic, but there are other treatises which may constitute Vedas—these are not published or translated from the original Sanskrit language. (Glover, 2014)

2.3 Writing Numbers—Digits and the Decimal Positional Place Value System

From this section on, we discuss only natural numbers $1, 2, 3, 4, \ldots$(Definition 1) together with zero.

Arithmetic is about performing arithmetic operations on natural numbers. Arithmetic operations are: addition, subtraction of a smaller number from a larger one, multiplication, and division—usually with remainder—this is what Division Algorithm in Section 1.2. is about.

Digits

Numerals are graphic representations of numbers. They are formed using basic symbols called digits.

The foundation of entire European Mathematics comes from India.

To do any computation, one has to be able to write (in general, visually represent—compare Chapter 3 "Inca," and Chapter 4 "Maya") numbers in a way which makes computation possible.

For this, Vedic Mathematics invented the positional place-value system of writing numbers. It has base ten and for this reason it is called decimal. The word "decimal" comes from the Latin word *decem* which means "ten." The decimal system of writing requires ten basic symbols—digits. The word "digit" comes from the Latin word *digitus* which means "finger." After all, humans have ten fingers and the fingers have been the basic tools for counting since time immemorial.

The shapes of the digits we use now are: 0, 1, 2, 3, 4, 5, 6, 7, 8, 9.

Figure 2.4 lists the names of the digits in Sanskrit.

The shapes of Sanskrit digits are listed in Figure 2.5:

Notice that the present symbols for 0, 2, and 3 are almost identical with Sanskrit symbols.

The most interesting and important digit is the one that represents, apparently, "nothing, " or, more precisely, absence, namely, 0. What does absence mean? Why a symbol for absence is needed? After all, it is easy to see 1 apple, 2 apples, and so on. But why the absence of apples has to be represented symbolically? The invention of O—the symbol for absence—was the greatest invention of humanity—not only in Mathematics, but in the context of the entire world civilization. This symbol O was invented in India. This is where the present 0 comes from. One name for zero in Sanskrit is shunya, which means "nothing."

In Chapter 4 about Maya arithmetic, you will find that the Maya had a symbol, in fact, several elaborate symbols, for the absence—zero. Hindu – Veda – Sanskrit had a very simple one, namely O, just a circle. The circle is considered a symbol of perfection, or, perhaps, even magic.

The most striking aspect of the symbol O for absence is another name for zero. In Sanskrit, this symbol O is also called pujya, which means "the most worshipful one." It must have been the most revered of all discoveries, almost God-like. This name suggests that the symbol O for absence really belongs to the realm of philosophy or mysticism. (Swamini, 2015)

0	shunya
1	eka
2	dvi
3	tri
4	chatur
5	pancha
6	shash
7	sapta
8	ashta
9	nava

Figure 2.4: Names of Sanskrit digits

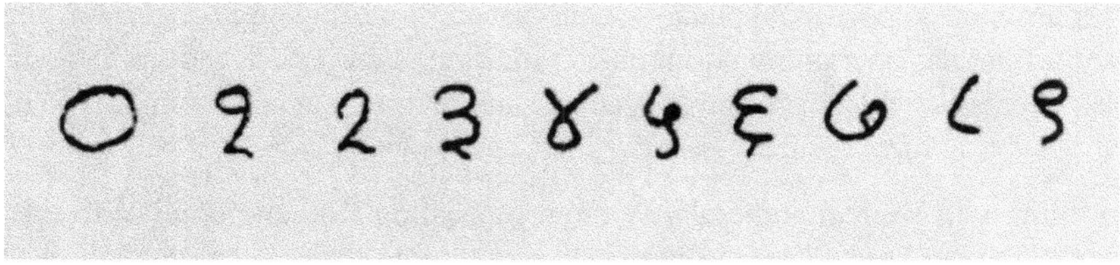

Figure 2.5: Shapes of Sanskrit digits, Szymanski, 2015

Positional place-value system

In the decimal system, that is, with the base ten (the symbol for ten is 10—see explanation below), each number is represented as the sum of the consecutive powers of 10 multiplied by the digits $0, 1, 2, 3, 4, 5, 6, 7, 8, 9$, starting from the right with 10^0. In this notation, we use Definition 14: $x^0 = 1$ for each number $x \neq 0$.

For example, $3 \times 10^2 + 6 \times 10^1 + 9 \times 10^0$ represents 3 hundreds 10^2 plus 6 tens 10^1 plus 9 units 10^0.

This number is written shortly as 369. This is why the system is called place-value system. The first digit from the right tells how many units are there in the number, the second from the right tells how many tens, the third from the right, how many hundreds, and so on. The place of the digit determines its value.

Let us now have a look at the number 639. It has 9 units, 3 tens, and 6 hundreds. Hence

$639 = 6 \times 10^2 + 3 \times 10^1 + 9 \times 10^0$. The digits used in 369 and 639 are the same, but their positions are different. This makes 369 and 639 different numbers. Therefore the location, position—place of the digit changes its value. This is why the full name of the system is positional place-value system.

The pictures—graphical representations—of numbers, like $369, 639$, are called numerals.

Notice that when we consider the polynomial, which we can write as: $6x^2 + 3x^1 + 9x^0$ (Definition 14 again), and replace x by 10, $x = 10$, we get 639.

This is how the subject of algebra came about—performing operations on letters rather than on numbers, as if letters were numbers, and then finding particular values by substituting numbers for letters as needed. In this sense, the concept of a polynomial can be seen as a generalization of the concept of a number. In the later section, Multiplication by **Urdhva-tiryak** Sutra, this idea plays the key role. The roots of algebra are in India, too.

Tobias Dantzig, 1930 —the author of the popular book written "for the cultured non-mathematician" expresses surprise that the ancient Greeks did not even think of algebra:

Particularly puzzling to us is the fact that the great mathematicians of Greece did not stumble on it. Is it that the Greeks had such marked contempt for applied science, leaving even the instruction of their children to the slaves? But if so, that the nation that gave us geometry, and carried this science so far, did not create even a rudimentary algebra? Is it not equally strange that algebra, that cornerstone of modern mathematics, also originated in India, and at about the same time that positional numeration did?

Now let us have a look why is 0 —the symbol for absence—so important. Suppose you count some items. You get 3 hundreds of items and 6 items. How do you write this in the decimal system? If you write 36, then you have just 3 tens and 6 units.

But you have 3 hundred and 6 units. Let us write it in the decimal system—we have to have consecutive powers of 10 starting with 10^0 from the right.

$3 \times 10^2 +$ *what?* $\times 10^1 + 6 \times 10^0$

It is clear that there are no tens—tens are absent. Fortunately, we have the symbol for absence, namely 0. Therefore *what?* is replaced by 0 and we get

$3 \times 10^2 + 0 \times 10^1 + 6 \times 10^0$

which we write shortly as 306.

When we write the number ten, we use the symbol 10, which means one ten and absence of units. Similarly, one hundred has symbol 100 meaning one hundred and absence of tens and units.

Without the symbol for absence no positional place—value system of writing numbers is possible. Without such system, it is very hard to perform arithmetic operations on numbers: addition, subtraction, multiplication, and division. Inca had such system which is also decimal—Chapter 3, and Maya had such system, but not decimal. The detailed description is in Chapter 4.

You are, certainly, familiar with Roman numerals. Without translating Roman numerals to the numerals we presently use, try to figure out how much is $C - LII$. And the question naturally arises: How does the Roman zero—symbol for absence—look like?

The famous French mathematician Pierre Simon, marquis de Laplace (1794–1827) said the following about the significance of the positional place-value number system:

It is India that gave us the ingenious method of expressing all numbers by the means of ten symbols, each symbol receiving a value of position as well as an absolute value, a profound and important idea which appears so simple to us now that we ignore its true merit, but its very simplicity, the great ease which it has lent to all computations, puts our arithmetic in the first rank of useful inventions, and we shall appreciate the grandeur of this achievement when we remember that it escaped the genius of Archimedes and Apollonius, two of the greatest minds produced by antiquity. (Hogben, 1967)

Indian numerals evolved over a period of centuries. The system which most closely resembles the present day notation for numerals and the use of the place-value system may be as old as the third century AD. Ever since ancient times calculation seems to have been performed in the decimal system. Vedic Mathematics seemed to be comfortable with very large numbers, for they had names for powers of ten up to 62. (Murty, 1995)

Yajurveda lists the following names for powers of 10 in Table 2.2 below.

The same list can be found in two places in the Taittriya Samhita.

For even larger numbers, the centesimal system—based on the base 100—was used.

The Latin word *centum* means "hundred," hence the name centesimal. From this Latin word, we have the English words: century, cent, percent (per hundred).

The Arabs got the digits and the positional place-value system from India. They brought it to Europe, because they traded between the East and Europe. They got familiar with the Mathematics of India and this is how the Indian Mathematics came to Europe. The Indian numerals in Europe were called Arabic, because the Europeans learned them and the positional place-value system from the Arabs. The Arabs use the name Hindsa for the ten digits, and **Hindsa** means: coming from Hind (short of Hindustan). India is called Hindustan, too, hence digits are coming from Hind. Presently, the numerals we use every day are called **Hindu–Arabic numerals**. (Datta, 2004)

Table 2.2: Powers of 10

		Sanskrit name
1	10^0	eka
10	10^1	dasha
100	10^2	shata
1,000	10^3	sahasra
10,000	10^4	ayuta
100,000	10^5	niyuta
1,000,000	10^6	prayuta
10,000,000	10^7	arbuda
100,000,000	10^8	nyarbuda
1,000,000,000	10^9	samudra
10,000,000,000	10^{10}	madhya
100,000,000,000	10^{11}	anta
1,000,000,000,000	10^{12}	praradha

Notation

In algebra, when a, b represent numbers then ab is the ordinary notation for their product, that is, the result of their multiplication (Definition 3): $ab = a \cdot b$

In the following discussion, the letters $a, b, c, \ldots$etc., $a_0, a_1, a_2, \ldots, b_0, b_1, b_2, \ldots$ will be used for digits $0, 1, 2, 3, 4, 5, 6, 7, 8, 9$.

To avoid confusion with multiplication, we introduce now a special notation to be able to write a number with digits represented by letters. This is necessary, because we will have to multiply one digit numbers, too.

For the product of digits a, b, we will write just ab, as usual. For the number with digits a, b, we will use a bracket and write $[ab]$. Thus $[ab]$ is the number:

$$[ab] = 10^1 a + 10^0 b = 10a + b.$$

The digit b is the number of units; the digit a is the number of tens. Similarly, the symbol $[abc]$ ($[a_2 a_1 a_0]$, resp.) stands for a three digit number with the units digit c (a_0, resp.), tens digit b (a_1,resp.), hundreds digit a (a_2, resp.), etc. This means

$$[abc] = 10^2 a + 10b + c, \; [a_3\, a_2\, a_1 a_0] = 10^3 a_3 + 10^2 a_2 + 10 a_1 + a_0$$

Since time immemorial people used numbers to count and to perform operations with them. Early Vedic Mathematics invented methods for fast computing. Addition and subtraction do not take much time. Multiplication and division are more complicated and require more time. Vedic sutras give hints how to perform multiplication and division more efficiently than what is being taught now in schools. We also find fast ways of squaring, cubing, and computing square and cube roots.

Multiplication

Multiplication was defined in Chapter 1 just before Definition 3.

Let us recall Definition 3.

a. **Factors** are numbers which are multiplied.

b. **The product** is the result of multiplication.

We learn multiplication of one digit numbers in early elementary school and memorize it as the so-called table of multiplication.

It is easy to do mental multiplication—that is, in your head—without any tools, including pencil and paper, when one number has one digit and the other has two digits. This is a very practical application of the distributive property:

$a(b + c) = ab + ac$ (Chapter 1, after Definition 3).

Here are two examples. More examples are in the exercises.

1. Multiply 8 and 47. We write 47 as $47 = 40 + 7$ and use the distributive property with $a = 8, \; b = 40, \; c = 7$. Hence we get:

$$8 \times 47 = 8 \times (40 + 7) = 8 \times 40 + 8 \times 7 = 320 + 56 = 376$$

2. $$6 \times 87 = 6 \times (80 + 7) = 6 \times 80 + 6 \times 7 = 480 + 42 = 522.$$

The idea is that all this can be easily done in your head.

When both numbers are two or more digit numbers, we usually multiply by hand on paper the way we learn in school.

There are, however, other methods of multiplying numbers, which are faster and more efficient than what we ordinarily learn in school. We find some of them in Hindu–Vedic Mathematics. One of them is presented here.

2.4 Multiplication using sutra 3 Urdhva–tiryak

Urdhva–tiryak means in Sanskrit "vertically and crosswise." The reason why this sutra is used will become clear later.

In Example 2.4.1, let us multiply two numbers with two digits each.

Example 2.4.1
Multiply 48 by 27. We first multiply the tens digits of both numbers $2 \times 4 = 8$. Then, multiply the tens digits of one number by the units digit of the other number and add the two products $4 \times 7 + 8 \times 2 = 44$. Finally, multiply the units digits of both numbers $8 \times 7 = 56$. Next, arrange the numbers as shown below.

	4	8		
	2	7		
	8	4	6	(a)
+	4	5		(b)
	12	9	6	

$$\begin{aligned} 4 \times 2 &= 8 \\ 4 \times 7 + 8 \times 2 &= 44 \\ 8 \times 7 &= 56 \end{aligned}$$

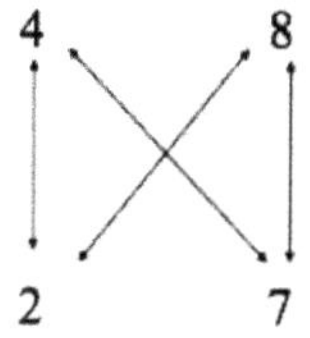

Notice how the numbers 8, 44, and 56 are aligned in the products on the left according to their decimal place value. In line (a), the number 8 is 4×2, the number 4 is the units digit of 44, and 6 is the units digit of 56. In line (b), the number 4 is the tens digit of 44, and 5 is the tens digit of 56. After adding the lines (a) and (b), we get the product 1296.

Thus, $48 \times 27 = 1296$

It is important to pay attention to the positional way of writing the steps leading to the answer.

To explain this method of multiplication, let us multiply the following two polynomials:

$(a_1x + a_0)(b_1x + b_0) = a_1b_1x^2 + (a_1b_0 + a_0b_1)x + a_0b_0$

(Chapter 1, Multiplication in 1.6).

If $x = 10$, then $a_1x + a_0$ is a two digit number in which a_0 is the units digit and a_1 is the tens digit. The same holds for the b's. The general scheme of the multiplication is depicted below.

		a_1	a_0	
		b_1	b_0	
	c_1	c_0	d_0	e_0
+		d_1	e_1	

$$\begin{aligned} a_1b_1 &= [c_1c_0] \\ a_1b_0 + a_0b_1 &= [d_1d_0] \\ a_0b_0 &= [e_1e_0] \end{aligned}$$

(In Example 2.4.1 $c_1 = 0$). This procedure can be graphically represented by the following easy to remember diagrams.

a_1b_1

$a_1b_0 + a_0b_1$

a_0b_0

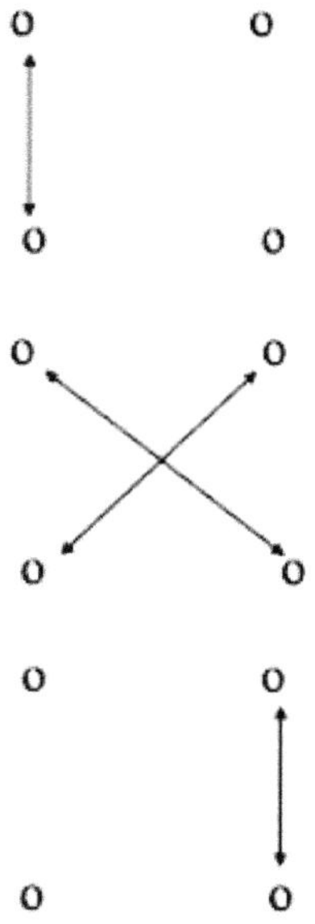

We can also put these three pictures together as follows.

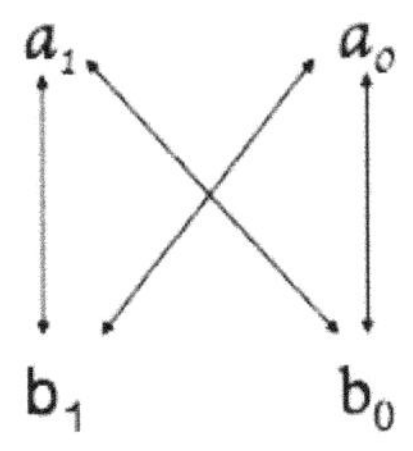

Multiply digits connected by line segments. Vertical segments are multiplications. The crosswise segments indicate adding the resulting products.

In Example 2.4.1, $[a_1\ a_0]=48$, $[b_1\ b_0]=27$, $a_1b_1=[c_1c_0]=08$, $a_1b_0+a_0b_1=[d_1d_0]=44$, $a_0b_0=[e_1e_0]=56$.

Example 2.4.1 and the explanation that follows is the simplest illustration of the principle contained in the Urdhva–tiryak sutra "vertically and crosswise" applied to a beautiful method of multiplying numbers. Notice that this method relies essentially on algebraic operation of multiplying polynomials.

It is clear that, really, multiplication of polynomials is based on Urdhva-tiryak sutra. (Chapter 1, Examples 1.6.6–1.6.12). For the numbers in the decimal system, we apply this method for the specific value of the variable $x=10$. This method works for numbers with any number of digits, as we are going to see later.

Here is one more example for the sake of clarity.

Example 2.4.2
Multiply 41 by 27.

	4	1		4×2	=	8
	2	7		$2\times 1+4\times 7$	=	30
	8	0	7	1×7	=	7 (think 07)
+	3					
	11	0	7			

Therefore, $41\times 27=1107$.

We can assume that both factors have the same number of digits. If not, then we can complete the "shorter" factor to match the length of the "longer" factor by inserting zeros at the beginning, for example, $142\times 26=142\times 026$.

To multiply three digit numbers, we will proceed in a similar way like in multiplication of two quadratic polynomials.

$$(a_2x^2+a_1x+a_0)(b_2x^2+b_1x+b_0)=$$
$$a_2b_2x^4+(a_2b_1+a_1b_2)x^3+(a_2b_0+a_1b_1+a_0b_2)x^2+(a_1b_0+a_0b_1)x+a_0b_0.$$

If $x=10$ the left hand side represent the product of two 3-digit numbers. We find the coefficients at x^4, x^3, x^2, x, and the free term, and proceed in a similar way as in the case of two digit numbers.

Example 2.4.3
Multiply 268×132. Here $[a_2a_1a_0]=268$, and $[b_2b_1b_0]=132$.

	2	6	8		
	1	3	2		
	2	2	0	6	6
+	1	3	3	1	
	3	5	3	7	6

$$
\begin{aligned}
a_2b_2 &= 2\times 1 &= 02\\
a_2b_1+a_1b_2 &= 2\times 3+6\times 1 &= 12\\
a_2b_0+a_1b_1+a_0b_2 &= 2\times 2+8\times 1+6\times 3 &= 30\\
a_1b_0+a_0b_1 &= 6\times 2+8\times 3 &= 36\\
a_0b_0 &= 8\times 2 &= 16\\
\text{Thus, } 268\times 132 &&= 35,376.
\end{aligned}
$$

Below you can see the diagrams which help remember the procedure.

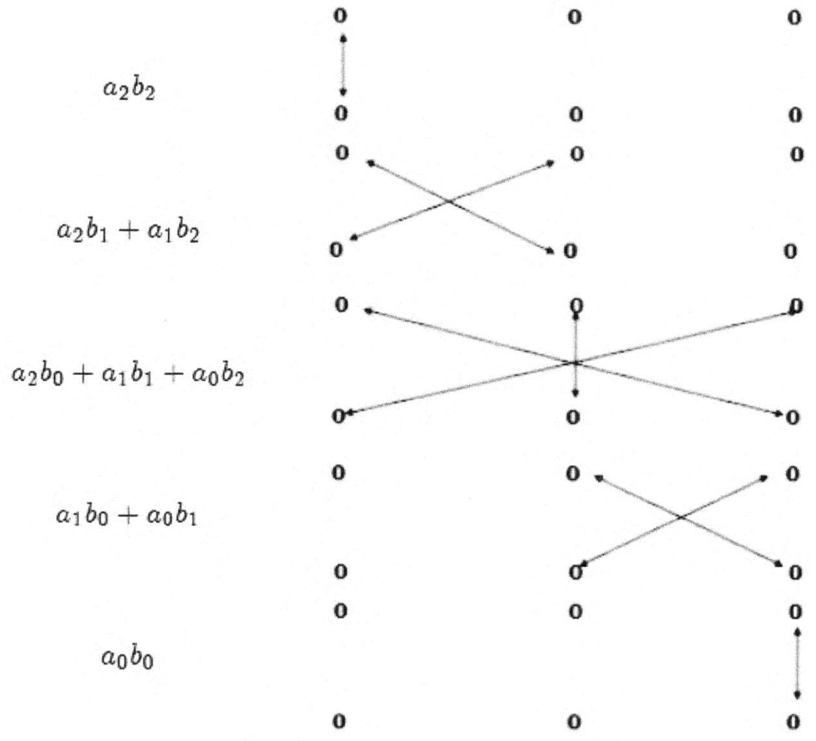

Example 2.4.4

Multiply 142 × 26. Here $[a_2a_1a_0] = 142$, $[b_2b_1b_0] = 026$.

	1	4	2	
	0	2	6	
	2	4	8	2
+	1	2	1	
	3	6	9	2

$$
\begin{aligned}
a_2b_2 &= 0\\
a_2b_1+a_1b_2 &= 2\\
a_2b_0+a_1b_1+a_0b_2 &= 14\\
a_1b_0+a_0b_1 &= 28\\
a_0b_0 &= 12
\end{aligned}
$$

In general, for multiplying n-digit numbers we get

$(a_nx^n + a_{n-1}x^{n-1} + ... + a_0)(b_nx^n + b_{n-1}x^{n-1} + ... + b_0)$ with $x = 10$.

Following the two particular previous cases, we find the coefficient at each x^k for $k = 0, \ldots n$. This coefficient is obtained by multiplying all $a_j b_{k-j}$, $j = 0, \ldots, k$, and forming the sum $\sum_{j=0}^{k} a_j b_{k-j} = a_0 b_k + a_1 b_{k-1} + a_2 b_{k-2} + \ldots + a_k b_0$.

Example 2.4.5
Multiply 2142 × 361. Here $[a_3 a_2 a_1 a_0] = 2142$, $[b_3 b_2 b_1 b_0] = 0361$

	2	1	4	2		
	0	3	6	1		
		6	5	0	1	6	2
+	1	2	3	1		
		7	7	3	2	6	2

$$
\begin{aligned}
a_3 b_3 &= 0 \\
a_3 b_2 + a_2 b_3 &= 6 \\
a_3 b_1 + a_2 b_2 + a_1 b_3 &= 15 \\
a_3 b_0 + a_2 b_1 + a_1 b_2 + a_0 b_3 &= 20 \\
a_2 b_0 + a_1 b_1 + a_0 b_2 &= 31 \\
a_1 b_0 + a_0 b_1 &= 16 \\
a_0 b_0 &= 2
\end{aligned}
$$

Here are the corresponding diagrams.

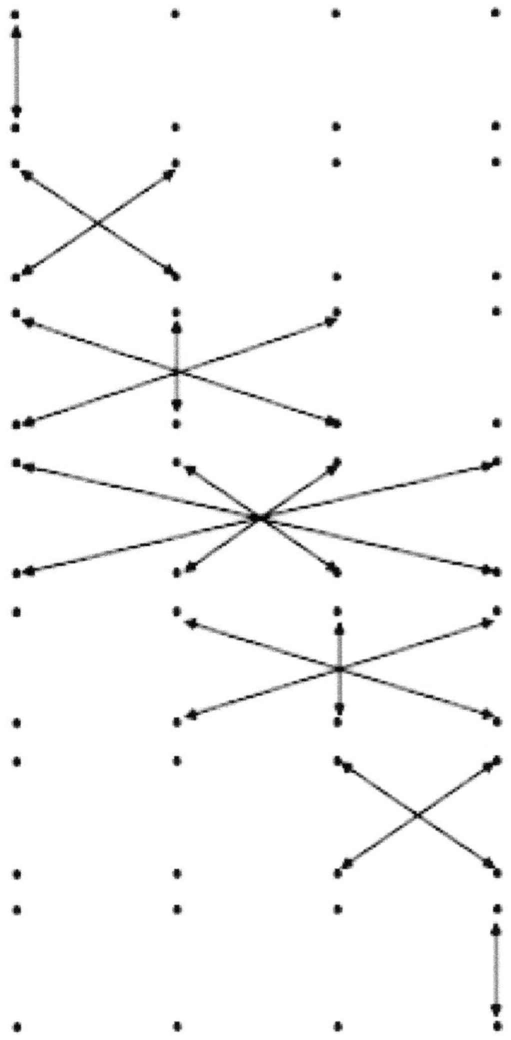

Notice the visual simplicity of the diagrams and that at each stage we are dealing with small numbers, no more than 3-digits, at least when the factors multiplied are of reasonable length (up to 5-digits, perhaps). With a little practice, this Vedic method of multiplication is considerable faster than the traditional method taught at schools. (Sri Bharati, 2010 & Gheverghese, 2011)

Squaring and Cubing

"Squaring" is another name for raising to the power 2. To square a number a, multiply it by itself twice: $a^2 = aa$, according to Definition 13.

The name "to square" comes from geometry. We define the area of a rectangle with sides of length a, b as ab, that is, the product of the numbers a, b, which is sometimes expressed as "length times width."

A square is a rectangle with all sides of equal length, $a = b$. Hence, the area of a square with sides of length a is $aa = a^2$.

Example 2.4.6

Square the number 237. Find 237 × 237. We follow Example 2.4.3.

$[a_2a_1a_0] = [b_2b_1b_0] = 237$

	2	3	7	
	2	3	7	
4	2	7	2	9
1	3	4	4	
5	6	1	6	9

Hence, $237^2 = 237 \times 237 = 56169$

When digits are not big, another shortcut is possible, for example to square 123, instead of multiplying the 3-digit number 123 by itself, we use 2-digit multiplication, but one of the "digits" is not a single digit but a group of digits. Simply, $[a_2a_1a_0] = 10[a_2a_1] + a_0$.

Example 2.4.7

Square the number 123.

		1	2	3	
		1	2	3	
	1	4	4	2	9
+			7		
	1	5	1	2	9

$$123 = 12 \times 10 + 3$$
$$a_1 = b_1 = 12, \text{ and } a_0 = b_0 = 3$$
$$a_1b_1 = a_1^2 = 144$$
$$a_1b_0 + a_0b_1 = 72$$
$$a_0b_0 = a_0^2 = 9$$

"Cubing" is another name for raising to the power 3. To cube a number a, multiply it by itself three times: $a^3 = aaa$, according to Definition 13.

The name "to cube" comes from geometry. We define the volume of a rectangular solid with sides of length a, b, c as abc, that is, the product of the numbers a, b, c, which is sometimes expressed as "length times width times height."

A cube is a rectangular solid with all sides of equal length, $a = b = c$. Hence, the volume of a cube with all sides of length a is $aaa = a^3$.

Cubing requires two multiplications. For a 2-digit number greater than 31, the square is a 4-digit number, so the second multiplication needed for cubing is longer. Vedic method uses the familiar formula for $(a+b)^3$ in the following way (the computation is in Example 1.6.10):

$$\begin{array}{lllllllllllll} (a+b)^3 & = & a^3 & + & & a^2b & + & & ab^2 & + & b^3 & & (1) \\ & & & + & 2 & a^2b & + & 2 & ab^2 & & & & (2) \\ \hline & & a^3 & + & 3 & a^2b & + & 3 & ab^2 & + & b^3 & & \end{array}$$

We added lines (1) and (2).

This method is, clearly, algebraic.

What makes this method faster than ordinary multiplication is that the numbers we deal with are smaller. Notice that the two terms of line (2) are twice the corresponding terms in line (1).

Example 2.4.8

Cube 26 using the formula above. Here $a = 20, b = 6$, so $a+b = 20+6 = 26, a^2b = 20^2 \times 6 = 2400,\ ab^2 = 20 \times 6^2 = 720$.

$$\begin{array}{ccccccc} a^3 & & a^2b & & ab^2 & & b^3 \\ 8000 & & 2400 & & 720 & & 216 \\ & & 2a^2b & & 2ab^2 & & \\ & & 4800 & & 1440 & & \\ \hline 8000 & + & 7200 & + & 2160 & + & 216 \end{array}$$

$$\begin{array}{lr} & 8000 \\ & 7200 \\ & 2160 \\ + & 216 \\ \hline & 17576 \end{array}$$

Hence, $26^3 = 17576$.

If we remember the decimal place value of the digits, we may write this computations schematically as follows:

$$\begin{array}{ccccc} & a^3 & a^2b & ab^2 & b^3 \\ & 8 & 24 & 72 & 216 \\ & & 2a^2b & 2ab^2 & \\ + & & 48 & 144 & \\ \hline & 8 & 72 & 216 & 216 \end{array}$$

Now the numbers in the last line have to be arranged according to the position of their digits in the decimal system, and added.

The alignment of digits in addition here and in the schematic version of Examples 2.4.8–2.4.10 is exactly the same as the alignment of digits in addition in Examples 2.4.1–2.4.7 as described in detail in Example 2.4.1.

$$\begin{array}{ccccc} & 10^3 & 10^2 & 10^1 & 10^0 \\ \hline & 8 & 2 & 6 & 6 \\ & 7 & 1 & 1 & \\ + & 2 & 2 & & \\ \hline & 17 & 5 & 7 & 6 \end{array}$$

Hence $26^3 = 17576$.

In case of a number with digits not too large, we can also cube 3-digit numbers treating a group of two digits as a single "digit" as in Example 2.4.6.

Example 2.4.9

Find the cube of 124. Here $a = 120$ and $b = 4$. Hence, $124 = 12 \times 10 + 4$, $a^3 = 1728000$, $a^2b = 120^2 \times 4 = 57600$, $ab^2 = 1920$, $b^3 = 64$.

a^3		a^2b		ab^2		b^3
1728000		57600		1920		64
		$2a^2b$		$2ab^2$		
		115200		3840		
1728000	+	172800	+	5760	+	64

	1728000
	172800
	5760
+	64
	1906624

Again, taking into account the decimal place values, we may write this computation schematically—without zeros at the end.

	a^3	a^2b	ab^2	b^3
	1728	576	192	64
		$2a^2b$	$2ab^2$	
+		1152	384	
	1728	1728	576	64

	17	2	8	8	6	4	
	1	7	2	7	6		
+			5				
	18	9	15	15	12	4	=
	19	0	6	6	2	4	

Hence 124^3 is 1906624.

Example 2.4.10

Find 6^6. We write $6^6 = (6^2)^3 = 36^3$. Here $a = 30$, and $b = 6$, $a^3 = 27000$, $a^2b = 540$, $ab^2 = 1080$, $b^3 = 216$.

$$
\begin{array}{ccccccc}
a^3 & & a^2b & & ab^2 & & b^3 \\
27000 & & 5400 & & 1080 & & 216 \\
 & & 2a^2b & & 2ab^2 & & \\
 & & 10800 & & 2160 & & \\
\hline
27000 & + & 16200 & + & 3240 & + & 216
\end{array}
$$

$$
\begin{array}{rr}
 & 27000 \\
 & 16200 \\
 & 3240 \\
+ & 216 \\
\hline
 & 46656
\end{array}
$$

If we skip zeros at the end, we can write this computation similarly to the way we wrote in the last two examples.

$$
\begin{array}{rcccc}
 & a^3 & a^2b & ab^2 & b^3 \\
 & 27 & 54 & 108 & 216 \\
+ & & 108 & 216 & \\
\hline
 & 27 & 162 & 324 & 216
\end{array}
$$

$$
\begin{array}{rccccc}
 & 2 & 7 & 2 & 4 & 6 \\
 & 1 & 6 & 2 & 1 & \\
+ & & 3 & 2 & & \\
\hline
 & 4 & 6 & 6 & 5 & 6
\end{array}
$$

Hence, 6^6 is 46656.

Division

In natural numbers, division is not an algebraic operation. The Division Algorithm is, in general, about division with remainder. We recall here the Division Algorithm from Chapter 1, with slightly different notation.

Division Algorithm

For any natural numbers d, p, there are unique natural numbers q, r such that

a. $d = pq + r$

b. $r < p$. Here q or r can be 0, but not both.

Note on terminology

In what follows, we will assume that the phrases: "p into d goes q times" or "we fit p into d, the result is q," have the same meaning.

The number d is called the dividend, p - the divisor, q - the quotient, r - the remainder. Using these names we restate the theorem:

a. dividend = divisor × quotient + remainder

b. The remainder is less than the divisor.

Example 2.4.11
If $d = 16,\ \ p = 5$, then $q = 3, r = 1$, and $16 = 5 \times 3 + 1$.

Division of any length dividend by a single-digit divisor is natural and can easily be done. If the divisor has more than one digit, we use long division which can be quite extensive if the dividend is long. It takes a considerable number of steps and space. The procedure starts with "fitting" the divisor p into a number formed by the first few digits of the dividend. This is the first digit q_1 of the quotient. Then, we multiply pq_1 and subtract from d remembering the proper decimal position. After bringing down the remaining digits of d the procedure repeats until r is less than p.

There are several division methods based on Vedic sutras. We will discuss one of them called "straight division," which is the simplest and shortest of them. The main difference between this method and long division is that we do not "fit" the whole divisor into a number formed from first digits of the dividend but only the first digit of the divisor. This is similar to long division of polynomials and will be explained later.

Example 2.4.12
Divide $d = 38982$ by $p = 73$.We set the table.

					24		29	0	(a)
3	3	8		9		8		2	(b)
7			3		3		1		(c)
		5		3		4			(d)

The divisor 73 is placed vertically in the leftmost column, the dividend is in line (b) (except the first 3), the quotient will appear in line (d), and the remainder is the last number in line (a).

Now we proceed to the first cycle.

- **i.** We look only at the first digit of p, that is, 7, and "fit" it into the first few digits of d, here, into 38. We get 5, which is the first digit of the quotient q, and place it in line (d) below 8 in line (b).
- **ii.** We multiply $7 \times 5 = 35$ and subtract $38 - 35 = 3$.We place that 3 in the line (c) between 8 and 9 of line (b).
- **iii.** Next, we look at the number 39 were 3 comes from the line (c), 9 from the line (b) next to 3 to the right. Then, multiply $3 \times 5 = 15$. Here, 3 is the units digit of the divisor p. Subtract $39 - 15 = 24$. Place 24 in line (a) between 9 and 8 of line (b). This is the full cycle of the procedure. Now we will repeat the cycle.

- **i.** We "fit" 7 into 24. The result is 3, which is the second digit of q. Place that 3 in line (d) under 9 in line (b).
- **ii.** $7 \times 3 = 21,\ 24 - 21 = 3$. Place 3 in line (c) between 9 and 8 of line (b).
- **iii.** Look at the number 38, where 3 comes from the line (c) and 8 from line (b) next to 3 to the right.
  $3 \times 3 = 9,\ 38 - 9 = 29$. Place 29 in line (a) between 8 and 2 of line (b).

Repeat the cycle again starting with 29.

- **i.** "Fit" 7 into 29. Get 4 which is the third digit of q.
- **ii.** $4 \times 7 = 28, 29 - 28 = 1$. Place 1 in line (c) between 8 and 2 of line (b).
- **iii.** Look at 12. Then multiply $4 \times 3 = 12$. Finally, $12 - 12 = 0 =$ the remainder.
  Hence, $q = 534,\ r = 0$. Thus, $38982 = 73 \times 534 + 0$.

The justification of this method has algebraic form and the resemblance of the long division of polynomials. Keep in mind that $x = 10$. Thus, $3x^3$ can be written as $3x^3 = 3 \times 10x^2$, or $3x^2 = 30x$. Hence, in the division below, $24x^2$ in line (β) is the result of the subtraction in line (α): $3x^3 - 6x^2 = 30x^2 - 6x^2 = 24x^2$. Similarly, $29x$ in line (δ) is the result of subtraction in line (γ): $3x^2 - x = 30x - x = 29x$.

$$
\begin{array}{rrrrrrrrrl}
 & 5x^2 & + & 3x & + & 4 & & & & \\
\hline
7x+3 \;) & 38x^3 & + & 9x^2 & + & 8x & + & 2 & & \\
 & 35x^3 & + & 15x^2 & & & & & & \\
\hline
 & 3x^3 & - & 6x^2 & & & & & & (\alpha) \\
 & & & 24x^2 & + & 8x & & & & (\beta) \\
 & & & 21x^2 & + & 9x & & & & \\
\hline
 & & & 3x^2 & - & x & & & & (\gamma) \\
 & & & & & 29x & + & 2 & & (\delta) \\
 & & & & & 28x & + & 12 & & \\
\hline
 & & & & & x & - & 10 & & \\
\hline
 & & & & & 10 & - & 10 & = 0 &
\end{array}
$$

Let us present the process in symbols.

$d = [d_4 d_3 d_2 d_1 d_0]$, $p = [p_1 p_0]$. We set the table as follows.

$$
\begin{array}{c|ccccccccc|l}
 & & & & & l_1 & & l_2 & r & & (a) \\
\hline
p_0 & d_4 & d_3 & & d_2 & & d_1 & & d_0 & & (b) \\
p_1 & & & k_1 & & k_2 & & k_3 & & & (c) \\
\hline
 & & q_2 & & q_1 & & q_0 & & & & (d)
\end{array}
$$

i. p_1 into $[d_4 d_3]$ goes q_2 times.
ii. $[d_4 d_3] - p_1 q_2 = k_1$.
iii. $[k_1 d_2] - p_0 q_2 = l_1$.

Repeat.

i. p_1 into l_1 goes q_1 times.
ii. $l_1 - p_1 q_1 = k_2$.
iii. $[k_2 d_1] - p_0 q_1 = l_2$.

Repeat again.

i. p_1 into l_2 goes q_0 times.
ii. $l_2 - p_1 q_0 = k_3$.
iii. $[k_3 d_0] - p_0 q_0 = r$.

We get $q = [q_2 q_1 q_0]$.

This is the procedure related to example 2.4.12. It is clear that it can be repeated any number of times depending on the number of digits of the dividend.

Sometimes, especially when the first digit of the divisor is small, extra anticipation and looking a step or two ahead is required because no negative numbers are allowed.

Example 2.4.13

Divide $d = 564291$ by $p = 27$. Let us start as in Example 2.4.12.

				2				
7	5		6		4	2	9	1
2		1		0				
	2		1					

i. 2 into 5 goes 2 times.
ii. $5 - 2 \times 2 = 1$.
iii. $16 - 7 \times 2 = 2$.

Repeat.

i. 2 into 2 goes 1 time. But if we choose 1 then in
ii. we get $2 - 2 \times 1 = 0$ and
iii. $04 - 7 \times 1 < 0$ which is not possible.

Therefore in (i) we choose 0 as the second digit of the quotient. This digit obtained by adjusting the fit in order to avoid negative numbers is underlined at the bottom line in the diagram below.

				2		24		26		26	18
7	5		6		4		2		9		1
2		1		2		8		8		8	
	2		$\underline{0}$		$\underline{8}$		$\underline{9}$		$\underline{9}$		

(ii) $2 - 2 \times 0 = 2$.

(iii) $24 - 7 \times 0 = 24$.

Repeat again.

i. 2 into 24 goes 12 times, but 12 is not a digit. By the experience in the previous step, we can easily guess that, actually, the third digit of the quotient is 8. Here, you have to see two steps ahead, why 9 is too big—explanation in parenthesis below.
ii. $24 - 2 \times 8 = 8$ $\quad$ $(24 - 2 \times 9 = 6)$.
iii. $82 - 7 \times 8 = 26$ $\quad$ $(62 - 7 \times 9 < 0)$.

And again,

i. 2 into 26 goes 13 times, which is not a digit. We pick 9 for the fourth digit of the quotient.
ii. $26 - 2 \times 9 = 8$.
iii. $89 - 7 \times 9 = 26$.

Finally, one more time,

i. 2 into 26 goes 9 times.
ii. $26 - 2 \times 9 = 8$.
iii. $81 - 7 \times 9 = 18$.

So $q = 20899$, $r = 18$, and $d = pq + r$, that is, $564291 = 27 \times 20899 + 18$.

It is an interesting exercise to write out this division using the "polynomial long division" like explanation, similar to the one after Example 2.4.12. This explanation precisely justifies choices of the digits of the quotient.

With experience and some practice this method of division is very fast and getting the digits of the quotient is completely natural.

Moreover, at each step we deal with small numbers, which can be easily dealt mentally.

Square roots

We will present a method for computing square roots based on Vedic sutras. For the sake of simplicity, we will work with small numbers. The resulting square roots will be typically 2-digit numbers $10a+b$. The method uses the algebraic formula (Chapter 1, Example 1.6.9):

$(10a+b)^2 = 100a^2 + 2 \times 10ab + b^2.$

Example 2.4.14

Find the square root of 529.

Consider the first digit of the given number: 5. Fit the largest possible square of a natural number into 5, in this case it is $2^2 = 4$. This is the first digit $a = 2$ of the square root. Multiply $4 \times 100 = 400$, subtract 400 from the original number to get $529 - 400 = 129$. Fit $20a = 40$ into 129 to get 3. This is the second digit $b = 3$ of the square root. Next, subtract $129 - 3 \times 40 = 9 = b^2$. These steps can be summarized below.

		5	2	9	Square root = 23
		4	0	0	
40)		1	2	9	
	−	1	2	0	
				9	
			−	9	
				0	

To justify the procedure let us look at the formula $(10a+b)^2 = 100a^2 + 2 \times 10ab + b^2$, which follows from Example 1.6.9. To begin, we find the first digit a of the square root by fitting the largest possible square of a natural number into the first digit (in the case of 3-digit numbers) or the first two digits (in the case of 4-digit numbers). Then we subtract $(10a+b)^2 - 100a^2 = 2 \times 10ab + b^2$. Next fit the number $2 \times 10a = 20a$ into the difference, this way we get the second digit b of the square root.We finish by subtracting $20ab$ and then b^2 to get 0.

Example 2.4.15

Find the square root of 225.

Consider the first digit of the given number: 2. Fit the largest possible square of a natural number into 2, in this case it is $1^2 = 1$. The first digit of the square root is $a = 1$. Multiply $1^2 \times 100 = 100$, subtract 100 from the original number, we get $225 - 100 = 125$. Fit $20a = 20$ into 125, we get 6. Instead of 6, we will use $b = 5$ because after subtracting $125 - 20ab$ we should get the square of a natural number, that is $125 - 100 = 25$, which is b^2.

		2	2	5	Square root = 15
		1	0	0	
20)		1	2	5	
	−	1	0	0	
			2	5	
		−	2	5	
				0	

Example 2.4.16
Find the square root of 1369.

Consider the first two digits of the given number: 13. Fit the largest possible square of a natural number into 13, in this case it is $3^2 = 9$. The first digit of the square root is $a = 3$. Multiply $9 \times 100 = 900$, subtract 900 from the original number, we get $1369 - 900 = 469$. Fit $20a = 60$ into 469, we get 7. This is the second digit $b = 7$ of the square root. Next, subtract $469 - 7 \times 60 = 49 = b^2$. The steps are summarized below.

	1	3	6	9		Square root = 37
	−	9	0	0		
60)		4	6	9		
		4	2	0		
			4	9		
		−	4	9		
				0		

In case the square root is a 3-digit number $[abc]$, the process of finding it is similar, but longer, because it is justified by the formula:

$$\text{(F)}\quad [abc]^2 = (10^2a + 10b + c)^2 = 10^4a^2 + 2 \times 10^3ab + 10^2b^2 + 2 \times 10(10a + b)c + c^2.$$

Example 2.4.17
Find the square root of 119716.

It is clear that it is a 3-digit number. The first digit is $a = 3$ because 3^2 is the largest square that fits into 11– the first two digits of 119716. Looking at formula (F) subtract $10^4a^2 = 10^43^2$ from 119716 to get 29716. Looking at (F) again it is clear that to obtain b we fit $2 \times 10^3a = 2 \times 10^3 \times 3 = 6000$ into 29716. The result is $b = 4$. Then, we subtract $29716 - 4 \times 6000 = 5716$. Next, we subtract $10^2b^2 = 1600$ from 5716 to get 4116, which is again justified by (F). To get c we fit $2 \times 10(10a + b) = 680$ into 4116. The result is $c = 6$. Finally, $4116 - 6 \times 680 = 36$, which is the square of c.

At first glance, the process may seem complicated, but (F) makes it clear how to proceed and why. At each stage, we find the digit and subtract its square. After subtraction, we perform a suitable division to get the next digit. Then the cycle repeats. The process can be clearly summarized in the following diagram:

		1	1	9	7	1	6	square root = 3 4 6
-10^4a^4			9					a b c
$2 \times 10^3a =$	6)		2	9				
-2×10^3ab			2	4				
				5	7			
-10^2b^2				1	6			
$2 \times 10(10a + b) =$	68)			4	1	1		
$-2 \times 10(10a + b)c$				4	0	8		
						3	6	
$-b^2$						3	6	
							0	

When keeping proper decimal positions of numbers, the zeros at the end can be dropped.

Cube roots

Finding the 2-digit cube root $[ab] = 10a + b$ of a number is based on the formula

(C) $[ab]^3 = (10a+b)^3 = 10^3a^3 + 3 \times (10a)^2b + 3 \times 10ab^2 + b^3$, which is a consequence of Example 1.6.10.

The procedure is similar to finding a square root.

Example 2.4.18

Find the cube root of 2197.

We begin by fitting the largest cube of a natural number into 2. This gives the first digit of the cube root: $a = 1$. By (C), we subtract $10^3a^3 = 1000$ from 2197 to get 1197. Then we fit $3 \times (10a)^2 = 300$ into 1197. The result is the second digit of the cube root, $b = 3$. Now we know that the cube root is 13. The rest is really a check. Looking at (C) we next subtract $3 \times (10a)^2b = 900$ from 1197 to get 297. Then we subtract $3 \times 10ab^2 = 270$ from 297 obtaining 27, which is the cube of $b = 3$.

Let us present the diagram illustrating this process.

		2	1	9	7	cube root = 1 3
-10^3a^3		1				
$3 \times (10a)^2 =$	3)	1	1			
$-3 \times (10a)^2b$			9			
			2	9		
$-3 \times 10ab^2$			2	7		
				2	7	
$-b^3$				2	7	
					0	

Like in the last example, the zeros and the end can be skipped if proper decimal positions are observed.

EXERCISES

1. Add mentally using sutra 10 Yavadunam - By the deficiency.

   a. $235 + 98$ **b.** $57 + 152$ **c.** $39 + 1234$
   d. $68 + 415$ **e.** $78 + 49$ **f.** $27 + 95$
   g. $47 + 92$ **h.** $96 + 77$ **i.** $48 + 336$
   j. $27 + 166$ **k.** $89 + 742$ **l.** $57 + 134$

2. Multiply mentally using the distributive property.

   a. 6×74 **b.** 7×36 **c.** 8×46
   d. 7×98 **e.** 6×57 **f.** 9×26
   g. 8×37 **h.** 4×76 **i.** 5×87
   j. 3×79 **k.** 9×68 **l.** 4×84

3. Multiply using sutra 3 Urdhva–tiryak—Vertically and cross-wise.

   a. 47×38 **b.** 27×63 **c.** 46×83
   d. 52×76 **e.** 84×26 **f.** 93×62
   g. 152×43 **h.** 274×76 **i.** 562×37

j. 724×46 **k.** 347×53 **l.** 626×47

m. 732×623 **n.** 374×653 **o.** 264×475

p. 1357×246 **q.** 3423×252 **r.** 5271×478

s. 2365×3527 **t.** 7326×2154 **u.** 3758×6743

4. Square the numbers using sutra 3 Urdhva–tiryak—Vertically and cross-wise.

a. 28^2 **b.** 76^2 **c.** 83^2

d. 253^2 **e.** 475^2 **f.** 647^2

g. 734^2 **h.** 321^2 **i.** 154^2

5. Square the numbers using the Vedic method of Example 2.4.7—consider the first two digits as a single "digit."

a. 112^2 $(112 = 11 \times 10 + 2)$

b. 115^2 $(115 = 11 \times 10 + 5)$

c. 118^2 $(118 = 11 \times 10 + 8)$

d. 124^2 $(124 = 12 \times 10 + 4)$

e. 125^2 $(125 = 12 \times 10 + 5)$

f. 134^2 $(134 = 13 \times 10 + 4)$

6. Cube the numbers using the Vedic algebraic method described in Example 2.4.8.

a. 34^3 **b.** 56^3 **c.** 72^3

d. 27^3 **e.** 63^3 **f.** 42^3

7. Cube the numbers using the Vedic algebraic method described in Example 2.4.9.

a. 112^3 **b.** 113^3 **c.** 117^3

d. 122^3 **e.** 126^3 **f.** 131^3

8. Find the following powers of numbers—use the method described in Example 2.4.10.

a. 2^6 **b.** 3^6 **c.** 4^6

d. 5^6 **e.** 7^6 **f.** 11^6

9. Given the dividend d and the divisor p find the quotient q and the remainder r using the method described in Examples 2.4.12 and 2.4.13

a. $d = 12345,\ p = 13$ **b.** $d = 24534,\ p = 22$ **c.** $d = 45341,\ p = 42$

d. $d = 53254,\ p = 17$ **e.** $d = 7243,\ p = 35$ **f.** $d = 11834,\ p = 51$

10. Find square roots of the following numbers using the Vedic method of Examples 2.4.14– 2.4.17.

a. 324 **b.** 676 **c.** 1225

d. 1764 **e.** 17956 **f.** 12544

11. Find cube roots of the following numbers using the Vedic method of Example 2.4.18.

a. 343 **b.** 1728 **c.** 3375

d. 12167 **e.** 29791 **f.** 46656

REFERENCES

Datta, Bibhutibhushan and Avadesh Narayan Singh. 2004. *History of Hindu Mathematics.* Delhi, India: Bharatiya Kala Prakashan.

Dantzig, Tobias. 1930. *Number: The Language of Science, (a critical survey written for the cultured non-mathematician).* New York: Macmillan.

Gheverghese, George. 2011. *The Crest of the Peacock: The Non-European Roots of Mathematics.* Princeton, NJ: Princeton University Press.

Glover, James. 2014. *Everything Vedic in Vedic Maths.* India: The Hindu.

Hogben, Lancelot. 1967. *Mathematics for the Million.* New York: W.W. Norton & Company Inc.

Sri Bharati Krsna Tirthaji Maharaja. 2010. *Vedic Mathematics.* Delhi, India: Motilal Banarasidass Publishers, Private Limited.

Malhotra, Rajiv and Aravindan Neelakandan. 2011. *Breakig India.* New Delhi, India: Amaryllis.

Murty, Kumar, V. 1995. *Contributions of Indian Subcontinent to Civilisation.* India: Prabuddha Bharata.

Swamini Svatmavidyananda. Private communication, 2015

RESOURCES

http://www.hinduwebsite.com/vedicsection/vedaindex.asp

http://www.thenagain.info/WebChron/India/India.html

http://www.thenagain.info/WebChron/India/Harappa.html

http://hindunet.org/hindu_history/ancient/aryan/aryan_frawley.html

CHAPTER 3

The Arithmetic of the Inca

To see them use another kind of quipu, with maize kernels, is a perfect joy. In order to carry out a very difficult computation for which an able computer would require paper and pen, these Indians make use of their kernels. They place one here, 3 somewhere else and 8, I know not where. They move one kernel here and there and the fact is that they are able to complete their computations without making the smallest mistake. As a matter of fact, they are better at practical arithmetic than we are with pen and ink... (de Acosta, 1590).

The Inca Empire arose in the central portion of the Andes and along the coastal plains to the west. The empire, which at its peak around 1525, extended from what is today Ecuador, Peru, parts of Bolivia, North and Central Chile, and Northwest Argentina was the greatest pre-Columbian empire of the Americas and existed only between ca. 1400 and 1533 AD. Its population was about 12 million at its apogee. The empire, which was in a civil war orchestrated by Huascar and Atahualpa who were half brothers fighting for power, collapsed after the arrival of the Spanish conquistadores led by Francisco Pizarro.

Quechua was the indigenous Andean language spoken by the Inca and is still spoken by many people of the Andes today. The word *Inca* means "ruler" in their Quechua language; later the Inca king was called *Sapa Inca* and the term Inca was used for the members of the noble class. The Quechua word *sapa* means "the only one." Eventually, the word Inca was applied as the name of the entire Empire and its people.

The Inca called their empire *Tawantinsuyu*, a Quechua word for "the land of four quarters." The word *tawa* means four in their language and *suyu* is the Quechua word for "quarter." We will see spoken numbers in Quechua later in this chapter.

The most influential pre-Columbian civilizations on the American continent were the Maya and the Inca. The Maya developed an elaborate writing system, part of their writings in the form of stone glyphs and codices are preserved until today. Therefore, we have "first hand" information about the Maya. The problem with the Inca is that they did not develop a writing system, hence, we do not have much information about the Inca from the Inca themselves. Most of the native information we have is preserved in the *quipus* (also written *khipus* or *quipos*) which have not yet been completely deciphered. Fortunately, the Spanish chroniclers left a significant amount of records about this civilization. A quipu is a recording device consisting of colored cords, hung from a main cord, containing information in the form of knots. The colored strings of cotton and sometimes camelid wool contained clusters of knots, in which the knot's placement represented units of 1, 10, 100, and so on, and the type of knot signified a number. The word *quipu* means "knot" in Quechua. It is believed that the color of the cord represented the item that was being reckoned—labor tributes, census, agricultural goods, livestock, textiles, ceramics, and a myriad of other items. A drawing of a portion of a quipu, found in Mollepampa in the northern part of Chile can be seen in Figure 3.1, right. The drawing on the left depicts a "blank quipu" in which the strings are ready to be knotted and attached to the primary cord, the quipu was found at Playa Miller in Arica, Chile. Both quipus are on display at Museo Chileno de Arte Precolombino.

It is believed that the quipus were somewhat used to convey written as well as numerical information. The numerical information has been studied extensively and many of the results have been published. Hopefully, in the future the written information will be decoded.

The *quipucamayoc* (*quipucamayu, khipucamayoc, khipukamayuq*), meaning "one who has charge of accounts," were a special tax-exempt class of citizens that were skilled in the recording of the quipus, as well as training new quipucamayocs. These individuals were educated from an early age in the quipucamayoc duties, and there was a hierarchy among them.

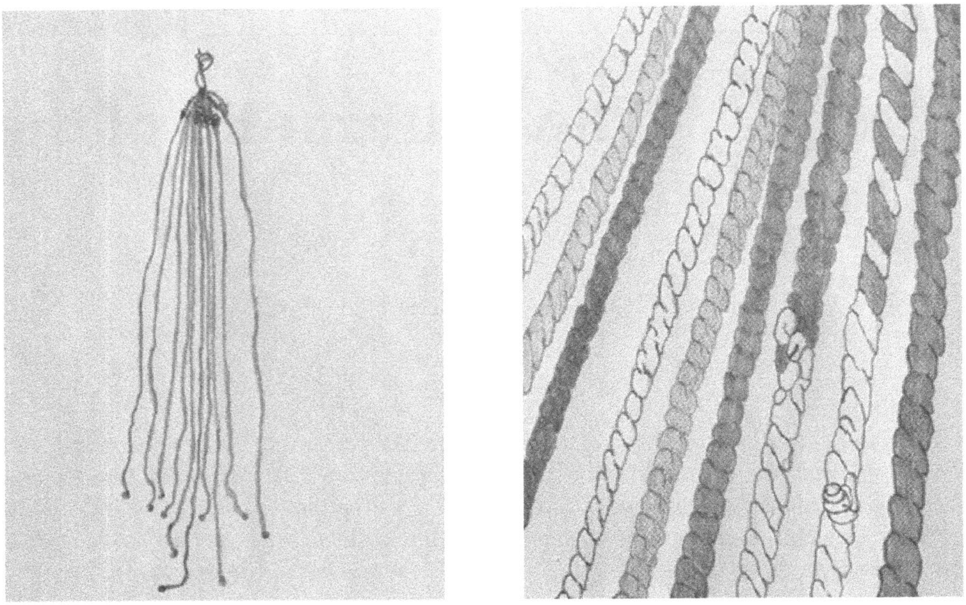

Figure 3.1: Quipus at Museo Chileno de Arte Precolombino, drawings by Ximena Catepillán

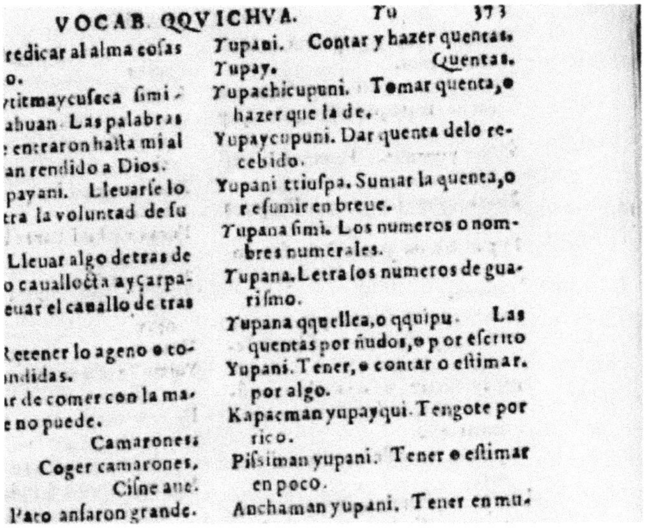

VOCAB. QQVICHVA. Yu 373

redicar al alma cosas
o.
ytitmaycusca simi-
ahuan. Las palabras
e entraron hasta mi al
an rendido a Dios.
payani. Lleuarse lo
tra la voluntad de su

Lleuar algo detras de
o cauallocta ayçarpa-
euar el cauallo de tras

Retener lo ageno o to-
ndidas.
r de comer con la ma-
e no puede.
Camarones.
Coger camarones.
Cisne aue.
Paco ansaron grande.

Yupani. Contar y hazer quentas.
Yupay. Quentas.
Yupachicupuni. Tomar quenta, o hazer que la de.
Yupaycupuni. Dar quenta delo recebido.
Yupani triuspa. Sumar la quenta, o resumir en breue.
Yupana simi. Los numeros o nombres numerales.
Yupana. Letra los numeros de guarismo.
Yupana qquellca, o qquipu. Las quentas por ñudos, o por escrito
Yupani. Tener, o contar o estimar. por algo.
Kapacman yupayqui. Tengote por rico.
Pissiman yupani. Tener o estimar en poco.
Anchaman yupani. Tener en mu-

From *Vocabvlario de la Lengva General de todo Perv llamada lengua Qquichua, o del Inca,* Lima, Peru, 1608.

Figure 3.2: From González Holguín (1989, page 373)

Moreover, every city, town, and village had these special accountants and treasurers of the Empire, and their number fluctuated between four and thirty. The quipucamayoc had to take account of the existence of products stored, such as corn, potatoes, beans, cotton, quinoa that farmers harvested. These quipu makers had to help surveyors to distribute land, tax collectors to control taxpayers, and astrologers to predict the planting and harvesting. In addition, they carried accounts of newborn children, marriages, deaths, among others. Apparently nothing escaped these skilled accountants, who were honest people, nonetheless at the time everyone had so little corruption in them that they might all be described as good.

However, in order to record the numerical information in the quipus, the quipucamayoc needed a number system. Hence, the Inca, like the Hindu-Arabic, developed a positional number system which was decimal. It is believed that the computations recorded on the quipus were performed in a grid-like stone, or wood, artifact called *yupana*. The word *yupana* in Quechua means "letter," the figures of numbers, and it is associated to *yupani* = reckoning, *yupai* = calculations, and *yupani triuspa* = summing up or making brief summaries (Figure 3.2). See González Holguín (1989, page 373).

Figure 3.3: The shaded region of the Inca Empire at its peak

The Inca Empire truly was an Empire—it was ruled by one emperor or king. A map of the Empire, which at its peak extended from what is today South Ecuador to Central Chile, is shown in Figure 3.3.

Following is the chronological Inca dynastic sequence. The time frames are only an approximation.

Manco Capac	1200s
Sinchi Roca	1230s
Lloque Yupanqui	1260s
Mayta Capac	1290s
Capac Yupanqui	1320s
Inca Roca	1350s
Yahuar Huacac	1380s
Viracocha Inca	1410–1438
Pachacuti Inca Yupanqui	1438–1471
Tupac Inca Yupanqui	1471–1493
Wayna Capac	1493–1527
Huascar	1527–1532
Atahualpa	1532–1533

Huascar was the eldest son of king Wayna Capac, but the sovereign seemed to favor Atahualpa, the youngest, who always accompanied him during his military campaigns. The smallpox epidemic that devastated the population upon the

Figure 3.4: Atahualpa is beheaded, Poma, (1613)

arrival of the Spanish conquistadores killed the king Wayna Capac and Huascar was designated the new ruler by the high priests. This started a malevolent civil war between the half brothers that caused thousands of deaths and culminated with the imprisonment and murder of Huascar by his own sibling Atahualpa.The last Inca ruler, Atahualpa, witnessed and participated in the main events leading to the destruction of his own empire which culminated with his brutal execution at the hands of Francisco Pizarro's army in July 1533. Guaman Poma's drawing (Figure 3.4) is a testimony of the despicable event that took place in the city of Cajamarca. The Inca chronicler Felipe Guaman Poma de Ayala (1535–1615) was one of the principal sources on information about the Inca. He was born into a noble Inca family shortly after the Spanish conquest of Peru, grew up in Cuzco and learned to read and write thanks to his mixed race half brother. Poma vividly recorded the events through a series of drawings with accompanying texts.

The capital of the Inca Empire, *Cuzco*, which is the Quechua word for "navel of the world," was located at its geographic center in the high Andes at 11,600 feet altitude. To rule successfully over such vast territory of over 300,000 square miles, an efficient system of communication was necessary. This was one reason why an extensive road system of approximately 14,000 miles was developed. The roads required the construction of multiple bridges of all sizes that were built using ropes. To make sure these bridges were secure enough they had to be renewed before the ropes became rotten, sometimes every six months (Figure 3.5). The chronicler Garcilaso de la Vega described in detail the construction of one of the bridges over the river Apurimac. The bridge was ordered to be built by one of the Inca kings to allow the passage of his army of about 12,000 warriors. Garcilaso de la Vega, born in Cuzco, was the son of a noble Spanish conqueror and an Inca princess, he was also a second cousin of the last two rulers of the Empire, Huascar and Atahualpa. He learned and witnessed some of the stories he wrote about while growing up among the Inca relatives of his mother and his father's military entourage. He learned Spanish and Latin while raised in his father's household absorbing both the Inca and Spanish traditions.

Countless earthquakes have hit the Andes since the Spanish conquest, but not a single Inca construction has fallen in more than 500 years. Their walls are trapezoidal in shape as they taper from a wide bottom to a narrower top. This provides additional stability during an earthquake. Some of the buildings have beds of smaller stones allowing the foundation to move slightly without harming the integrity of the walls (Barnhart, 2012). Moreover, they built a series of terraces and channels on the mountains and hillsides for agricultural purposes. The efficient system required the construction of aqueducts some of which are still in use. All these facts, among others, corroborate that the Inca were advanced engineers, and engineering, in turn, requires computations beginning with arithmetic (See pictures by Ximena Catepillán, 2015, in Figure 3.6).

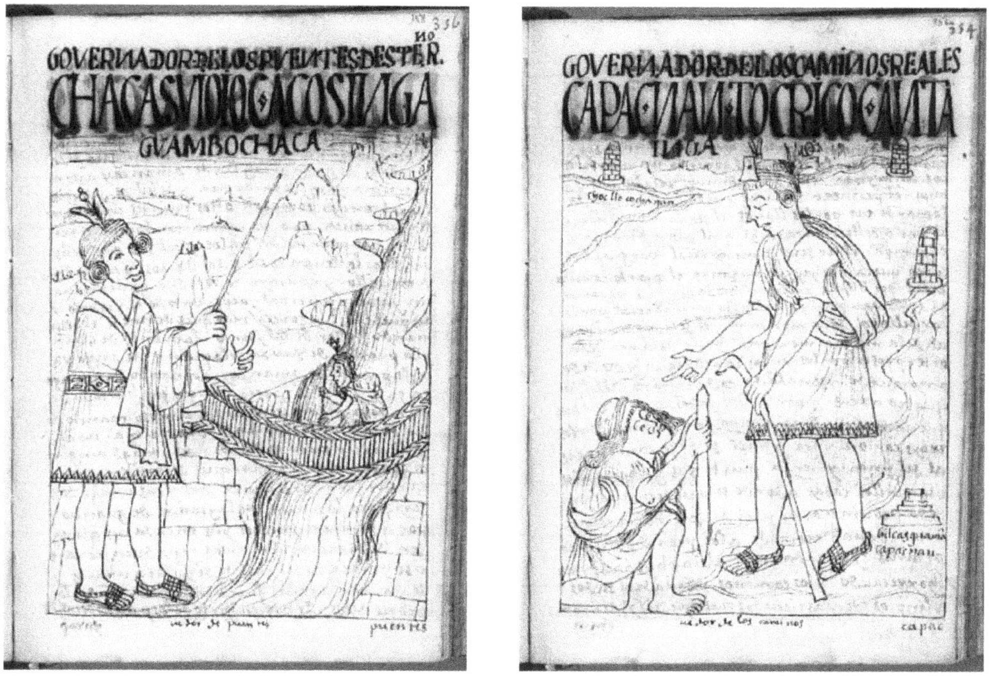

Figure 3.5: An overseer of the Inca bridges (left) and highways (right), Poma, (1613)

Figure 3.6: Wall details at Ollantaytambo, left and Machu Picchu, right. (Catepillán, 2015)

Guaman Poma also drew illustrations of philosophers and astrologers whose duties were, among others, the counting of time: From one day to ten days they counted a week, then thirty, thirty one, or thirty two days for a month according to the moon. They had names for each month, and the year was divided into twelve months by the moons. This explains why they used the same Quechua word *quilla* for "month" and "moon." They had special ceremonies, sacrifices, and activities for each month. Since the observation of lunar phases to divide the year quickly loses synchrony with a solar year, they used the solstices to keep the year in sync with the seasons for agricultural purposes. There is evidence that they used a twelve month calendar, see Table 3.1, that began with the first new moon including or following a solstice (Bauer and Dearborn, 1995).

Several chroniclers have written about the evidence of Inca astronomical observations by the use of pillars along the Cuzco horizon to track the solar motion. In the next paragraph, you will read chronicler Cieza de Leon's quote.

In another section stands Karmenka hill, where at intervals there are small towers which they used to study the movement of the sun, to which they attached great importance.

Table 3.1: The twelve-month Inca calendar

Month		Inca Month
January (23 December)	The day after the solstice	Capac Raymi Camay Quilla
February (22 January)		Paucar Uaray Hatun Pocoy Quilla
March (21 February)		Pacha Pocuy Quilla
April (23 March)	Approx. two days after the equinox	Ynca Raymi Mamay Quilla
May (22 April)		Atun Cusqui Aymoray Quilla
June (23 May)		Haucay Cusqui Quilla
July (23 June)	The day after the solstice	Chacra Conacuy Quilla
August (23 July)		Chacra Yapuy Quilla
September (23 August)		Coya Raymi Quilla
October (22 September)	Two days before the equinox	Uma Raymi Quilla
November (22 October)		Aya Marcay Quilla
December (22 November)		Capac Ynti Raymi

The hill mentioned in Cieza de Leon's quote is in the northwest part of Cuzco. He also wrote that the Inca could tell by the shadow the sun made on the pillars when to plant and other things they gathered from the information. We know that the number of pillars was at least four but the exact number is unknown as none of the chroniclers that wrote about them gave a specific number.

A list of the Inca months according to Guaman Poma can be seen in Table 3.1, there are several other lists from other chroniclers as well (Bauer and Dearborn, 1995).

The Inca followed with great interest the movements of the *Collca*, the Quechua name for the Pleiades. *Chacana*, their name for the belt of Orion, was also studied by them among other stars. These people described the milky way to the chronicler Garcilaso de la Vega as the figure of an ewe with the body of a suckling lamb.

In what follows, we introduce the spoken numbers in Quechua and continue with theories concerning computations with the yupana. We illustrate the uses of the yupana with examples.

3.1 The Number System of the Inca

Spoken Numbers in Quechua

Unlike Hindus or Maya, the Inca did not have written numerals or digits. This is why spoken numbers in Quechua are so important. We begin listing the Quechua names of the basic numbers (Urton, 1997), in base ten number system. Table 3.2. To write compound numbers, the system follows addition and multiplication rules. In Quechua, the words *yapay* and *miray* are used for addition and multiplication, respectively.

The Quechua term *hunu* (huno) is described as the number 10,000 although Urton (1997), did not hear the word among the peoples of the villages of Peru and Bolivia during his research. There is no evidence of a Quechua name or symbol for the number zero although they represented it in the quipus by the absence of a knot. The Inca did not have written numerals (pictures), hence, the symbol for zero was not there, either. The absence is exactly what the symbol for zero represents. For the concept of infinity they sometimes used the words *waranqa waranqa*, which correspond to 1 million as we will see later, and, *mana yupayta atina* for "what cannot be counted." Perhaps the words *pantacac hunu (huno)* at the end of page 361 (Poma) translate as "uncountable."

Table 3.2: The Quechua names for numbers*

Quechua name	Number
uj	one
iskay	two
kinsa	three
tawa	four
phishqa	five
suqta	six
qanchis	seven
pusaq	eight
jisqon	nine
chunka	ten
pachaq	one hundred
waranqa	one thousand
hunu	ten thousand

... y contava des tamanera
comensando de uno dos y tres
suc - yscay - quinza - taua -
pichica - zocta - canchis - puzac
- yscon - chunga - yscaychunga
- quinzachunga - tauachonga -
piscachunga - zoctachunga -
canchischunga pozacchunga -
ysconchunga - pachaca uaranga
chungauaranga - huno -
pachacahuno - uarangahuno -
pantacachuno ______.

* From *Nueva Coronica y Buen Gobierno* by Felipe Guaman Poma de Ayala, 1613.

Figure 3.7: Poma's description of the Quechua numbers, 1613*

Page 361 of the manuscript "Nueva Coronica y Buen Gobierno" written by Felipe Guaman Poma de Ayala (1613), is devoted to Inca numbers. In Figure 3.7, we can see the paragraph describing the numbers.

How to write (speak) numbers

The rules described below are used in the formation of compound numbers. In rules 1–4, we use a basic addition principle while in rule 5, we utilize the multiplication principle and in rule 6, we combine the operations of addition and multiplication.

Note: In addition the higher number precedes the lower number whereas in multiplication the lower number precedes the higher number.

Rule 1
When the last digit ends in a vowel we write the suffix **yuq** at the end, which indicates addition.

$13 = 10 + 3 =$ chunka kinsa**yuq** = 10 (chunka) + 3 (kinsa)

$114 = 100 + 10 + 4 =$ pachaq chunka tawa**yuq** = 100 (pachaq) + 10 (chunka) + 4 (tawa)

$1,005 = 1,000 + 5 =$ waranqa phishqa**yuq** = $1,000$ (waranqa) + 5 (phishqa)

$1,114 = 1,000 + 100 + 10 + 4 =$ waranqa pachaq chunka tawa**yuq** = $1,000$ (waranqa) + 100 (pachaq) + 10 (chunka) + 4 (tawa)

$10,116 = 10,000 + 100 + 10 + 6 =$ hunu pachaq chunka suqta**yuq**

Rule 2
When the last digit ends in a consonant or semiconsonant, the syllable **ni** precedes the suffix **yuq,** hence, we write **niyuq** at the end, which indicates addition.

$11 = 10 + 1 =$ chunka uj**niyuq** = 10 (chunka) + 1 (uj)

$12 = 10 + 2 =$ chunka iskay**niyuq** = 10 (chunka) + 2 (iskay)

$1,108 = 1,000 + 100 + 8 =$ waranqa pachaq pusaq**niyuq** = $1,000$ (waranqa) + 100 (pachaq) + 8 (pusaq)

$1,119 = 1,000 + 100 + 10 + 9 =$ waranqa pachaq chunka jisqo**niyuq** = $1,000$ (waranqa) + 100 (pachaq) + 10 (chunka) + 9 (jisqon)

$11,007 = 10,000 + 1,000 + 7 =$ hunu waranqa qanchisni**yuq**

Rule 3
If a three or more digit number ends with 10 (chunka), we write **chunkan** instead of chunkayuq.

$110 = 100 + 10 =$ pachaq chunkan

$1,010 = 1,000 + 10 =$ waranqa chunkan

$1,110 = 1,000 + 100 + 10 =$ waranqa pachaq chunkan

$10,020 = 10,000 + 20 =$ hunu iskay chunkan

Rule 4
When the last digit in an addition is pachaq, we write **pachaqnin** instead of pachaqniyuq, for example,

$1,100 = 1,000 + 100 =$ waranqa pachaqnin

$10,100 = 10,000 + 100 =$ hunu pachaqnin

$11,100 = 10,000 + 1,000 + 100 =$ hunu waranqa pachaqnin

The next rule, is a multiplication rule.

Rule 5
To write multiples of powers of ten we simply write the number followed by tens, hundreds, thousands, ten thousands, hundred thousands, and so on. The lower number precedes the higher number. What is being used in this rule is a basic multiplication principle. See examples in the following.

$10,000$ = ten thousands = $10 \times 1,000$ = chunka waranqa = hunu

100, 000 = one hundred thousands = 100 × 1, 000 = pachaq waranqa

1, 000, 000 = one thousand thousands = one million = 1, 000 × 1, 000 = waranqa waranqa = 100 × 10, 000 = pachaq hunu

10, 000, 000 = ten thousand thousands = 10 × 1, 000 × 1, 000 = chunka waranqa waranqa = waranqa hunu

30 = 3 tens = 3 × 10 = kinsa chunkan

300 = 3 hundreds = 3 × 100 = kinsa pachaqnin

3, 000 = 3 thousands = 3 × 1, 000 = kinsa waranqa

30, 000 = 3 ten thousands = 3 × 10 × 1, 000 = kinsa chunka waranqa = kinsa hunu

300, 000 = 3 hundred thousands = 3 × 100 × 1, 000 = kinsa pachaq waranqa

3, 000, 000 = 3 thousand thousands = 3 × 1, 000 × 1, 000 = kinsa waranqa waranqa

30, 000, 000 = 3 ten thousand thousands = 3 × 10 × 1, 000 × 1, 000 = kinsa chunka waranqa waranqa = kinsa waranqa hunu

Rule 6

In this rule we combine the addition and multiplication rules. Here, the multiplication is performed before addition, as we do in algebra. Remember that the lower number precedes the higher number in multiplication, and, in addition the higher number precedes the lower number.

48 = 40 + 8 = (4 × 10) + 8 = (tawa chunka) + (pusaq) = tawa chunka pusaq**niyuq**

130 = 100 + 3 tens = pachaq kinsa chunkan

3, 300 = 3 thousands + 3 hundreds =kinsa waranqa kinsa pachaqnin

5, 008 = 5 thousands + 8 = phishqa waranka pusaq**niyuq** = 5, 000 (phishqa waranka) + 8 (pusaq)

3603 = 3 thousands + 6 hundreds + 3 = kinsa waranqa suqta pachaq kinsa**yuq** = 3, 000 (kinsa waranqa) + 600 (suqta pachaq) + 3 (kinsa)

66, 000 = 66 thousands = suqta chunka suqtayuk waranqa = [60 (suqta chunka) + 6 (suqta)] × 1, 000 (waranqa)

2, 345 = 2 thousands + 3 hundreds + 4 tens + 5 = iskay waranqa kinsa pachaq tawa chunka phishqa**yuq**

88, 888 = 8 ten thousands + 8 thousands + 8 hundreds + 8 tens + 8 = pusaq hunu pusaq waranqa pusaq pachaq pusaq chunka pusaq**niyuk**

Notice that in various cases the numbers can be written in more than one combination of words.

3.2 The Yupana and Quipu

A yupana is a rectangular grid, usually on a piece of stone, clay, or wood, in which a collection of corn kernels (or beans) of different colors are used to represent numbers on the grid. Corn and beans were a fundamental staple for the Inca and had many varieties of colors. Poma mentioned also the use of quinoa to depict the numbers. This might sound strange because of the small size of the quinoa grains, and perhaps the somewhat larger "pearled" quinoa grains were used. In fact, some yupanas had very small cells.

Evidently when performing computations the Inca had sometimes an excessive amount of kernels, and it is believed that the use of different colors was used to overcome the problem. For example, they might have replaced ten yellow kernels (beans) by a single brown kernel (bean) or similar.

There is also evidence that they used a board grid to play the *taptana*, a game that was played with a pyramidal shape dice. The Quechua word *taptana* was translated as "alquerque game," the strategy board game, by González Holguín (page 338).

Courtesy of Ximena Catepillan

Figure 3.8: Yupana, drawing by Catepillán, (2015)

Figure 3.8 shows the drawing of a stone yupana found in the Áncash region in Perú. (Radicati Di Primeglio, 1973)

Yupanas were used to make computations, the results of which were recorded on a quipu. The reader will agree that it is not difficult to perform operations on the yupana and, in fact, is much easier and faster than traditional style on paper—see quote by Father Jose de Acosta below from his book "Historia Natural Moral de las Indias" (1940 ed.). In the passage Acosta, a Spanish priest, who lived in Peru from 1571 to 1586, describes the use of a yupana.

To see them use another kind of quipu, with maize kernels, is a perfect joy. In order to carry out a very difficult computation for which an able computer would require paper and pen, these Indians make use of their kernels. They place one here, 3 somewhere else and 8, I know not where. They move one kernel here and there and the fact is that they are able to complete their computations without making the smallest mistake. As a matter of fact, they are better at practical arithmetic than we are with pen and ink...

To reinforce the description of Acosta, we quote the following passage from Inca Garcilaso de la Vega (2004 ed.), who was the first Peruvian Inca/Spanish chronicler (1539–1616):

De la Aritmetica supieron mucho y por admirable manera, que por nudos dados en unos hilos de diversos colores daban cuenta de todo lo que en el reino del Inca había de tributos y contribuciones por cargo y descargo. Sumaban, restaban y multiplicaban por aquellos nudos, y, para saber lo que cabia a cada pueblo, hacían las particiones con granos de maíz y piedrezuelas, de manera que les salía cierta su cuenta...

Of Arithmetic they knew a lot and in an admirable manner, since by using knots on different color threads they kept track of everything that in the kingdom of the Inca had to do with taxes and dues by adding and subtracting. They added, subtracted, and multiplied using knots to keep track of each town's dues, they did the computations using kernels of corn and pebbles to obtain accurate results...

The Poma's drawing (Figure 3.9) depicts a *Tawantin Suyu*—an accountant in Quechua—who is holding a quipu and, in the lower left corner, the grid resembles a yupana with dots representing numerical information. The text on the picture translates as follows: Main Accountant and Treasurer, *Tawantin Suyu khipuq kuraka*, "the authority in charge of the quipus of the Tawantinsuyu." The picture is not the only one in Poma's manuscripts where similar accountants, treasures, and secretaries are portrayed with quipus in their hands.

For brevity, in what follows we will use the word "grid" for yupana and "dot" for a black dot representing a corn kernel or bean (Figure 3.10). The empty circles represent places for dots.

Figure 3.9: Drawing by Poma, (1613)

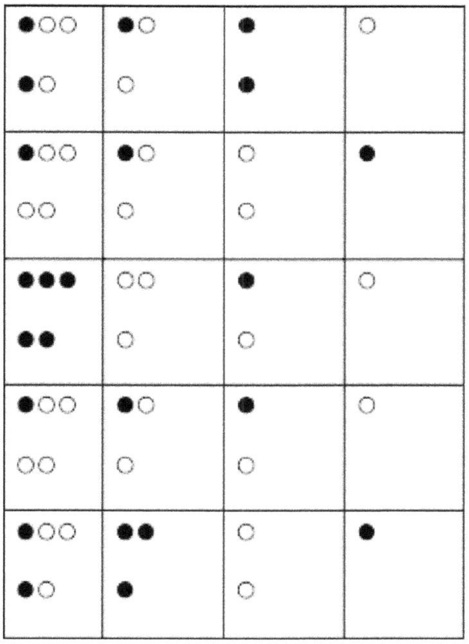

From *Nueva Coronica y Buen Gobierno* by Felipe Guaman Poma de Ayala, 1613.

Figure 3.10: The grid in Poma's drawing, 1613

Let us have a look at the grid of Poma's drawing:

Wassen (1931) gives the following interpretation of this grid:

1. Row values, from bottom to top, are successive powers of ten.
2. Column values, from left to right, are the values 1, 5, 15, and 30.

The consecutive rows (from bottom to top) represent the following numbers:

Row 1: $2 \times 1 + 3 \times 5 + 1 \times 30 = 47$

Row 2: $(1 + 5 + 15) \times 10 = 210$

Row 3: $(5 + 15) \times 100 = 2,000$

Row 4: $(1 + 5 + 30) \times 1,000 = 36,000$

Row 5: $(2 + 5 + 30) \times 10,000 = 370,000$

When we add all these numbers we get 408, 257.

A much more natural, purely decimal, explanation was proposed by Gheverghese (1990).

Row values, from bottom to top are successive powers of ten. Column values, from left to right, have the value 1.

Therefore, Poma's drawing depicts now the number

$6 + 3 \times 10 + 6 \times 100 + 3 \times 1,000 + 5 \times 10,000 = 53,636$

In what follows we will only use the decimal system of Gheverghese. Before describing the arithmetic operations on a yupana we will illustrate in the following figure how naturally a numerical representation on a yupana can be knotted on a quipu. The figure is consistent with Father Jose de Acosta's statement "*...su modo no era escribir reglon seguido, sino de alto abajo...* " that translates *"...their writing method was not line by line but from top to bottom...."*

In Figure 3.11, we read the numbers 20,311 left, 10,243 middle, and 1,010 right and (from top to bottom) in both: the grid on the left and the quipu on the right.

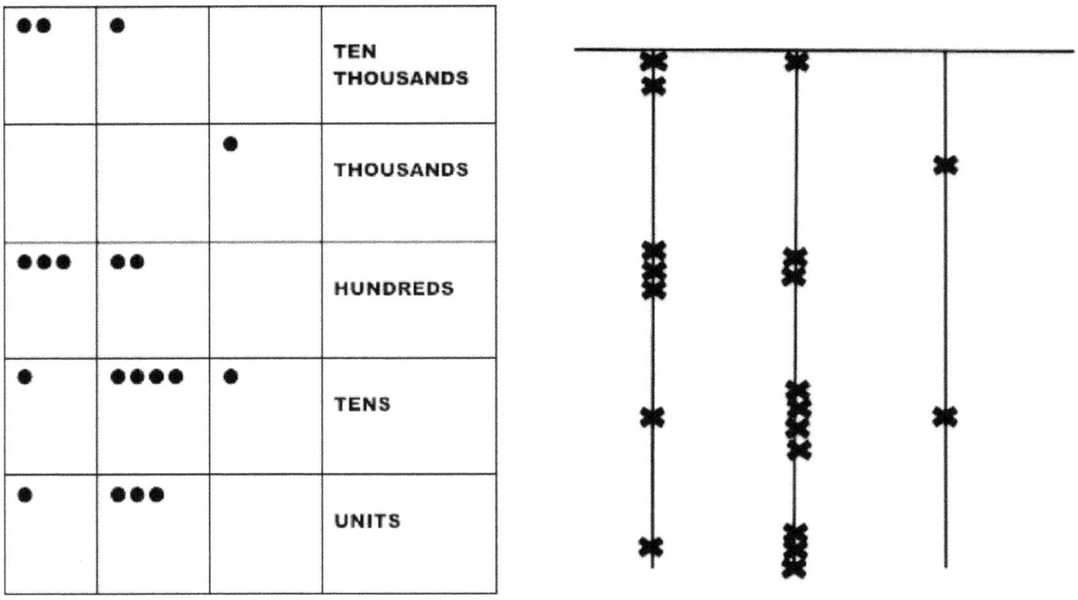

Figure 3.11: The yupana on the left and the quipu on the right

Now we shall discuss arithmetic operations performed using the grid. Addition, subtraction, and multiplication follow the interpretation of Radicati Di Primeglio (1973). Our own interpretation will be used to describe division. Examples will be used to illustrate the arithmetic operations, which will be performed using natural numbers 1, 2, 3, ..., and 0.

3.3 Arithmetic Operations on the Yupana

Addition on the Yupana

Example 3.3.1

To add the numbers 21,512 and 11,013 we place them in the first two left columns of the grid. In each row, we move the dots from the first two cells into the last cell. In the fourth column, we get the answer 32,525. Keep in mind that all these operations were performed manually by moving dots (kernels) from one cell of the yupana to another.

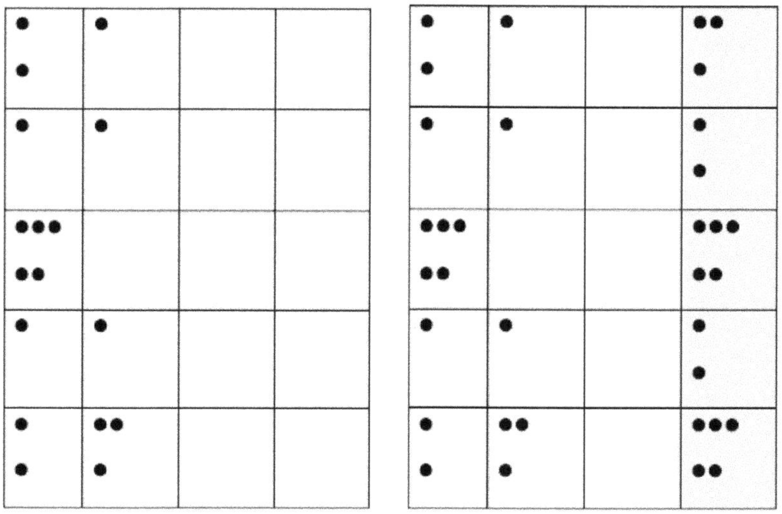

Example 3.3.2

Let us add now 31,511 and 1,652. We obtain 3, 2, 11, 6, and 3 dots in the fourth column (from top to bottom). We have to trade since we cannot have ten or more dots in any cell. Therefore, we rewrite the last column with 3, 3, 1, 6, 3 dots, which means that the result is 33, 163. This is because every ten dots in any cell can be replaced by one dot in the cell right above it (decimal system).

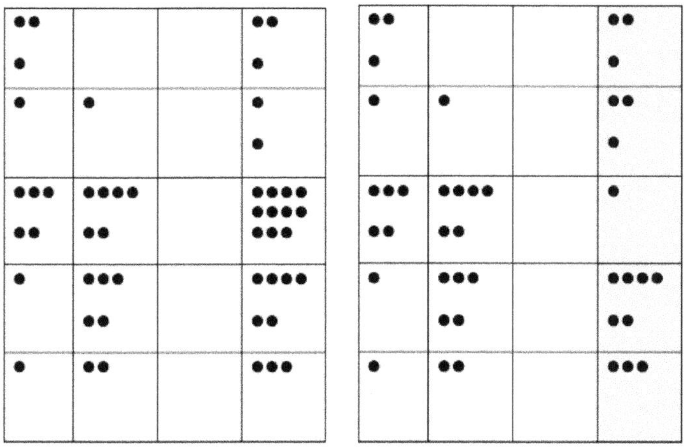

The trading rule is: Ten dots in a cell equals one dot in the cell directly above.

Subtraction on the Yupana

Subtraction without trading is straightforward.

Example 3.3.3

To subtract 532 from 2, 341 we have to trade. We use the trade rule to rewrite the first column as 1, 13, 3, 11 and then subtract obtaining 1, 809 in the last column.

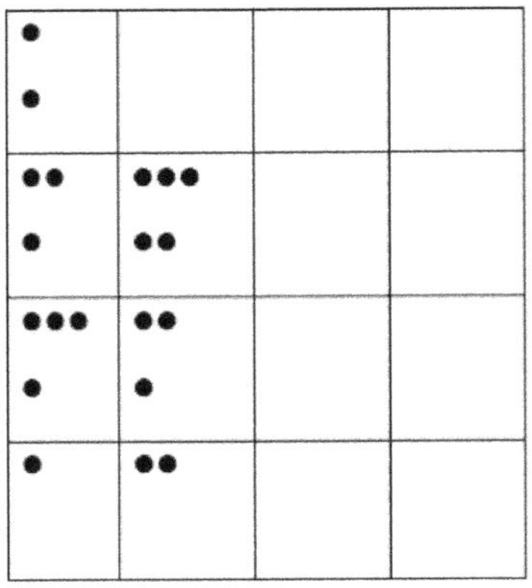 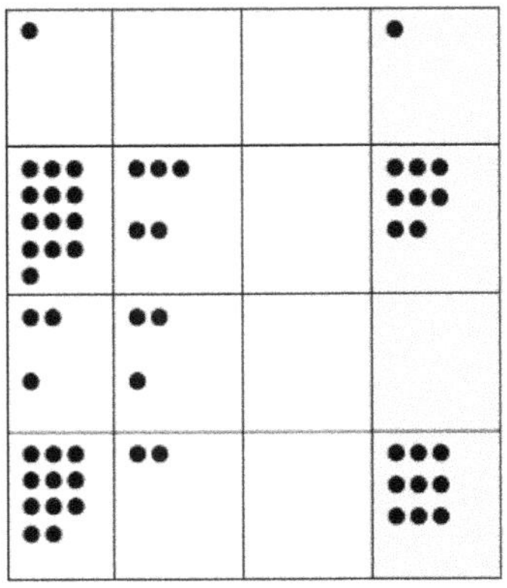

Multiplication on the Yupana

To multiply numbers, Radicati Di Primeglio (1973) was inspired by the multiplication method used by Calderon (1966) for the vigesimal Maya number system.

Example 3.3.4

Multiply 23 by 211. Place one of the factors, 23, in our example

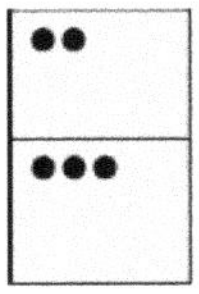

in the first column from the left and the second factor, 211, horizontally in the first row starting with the second column

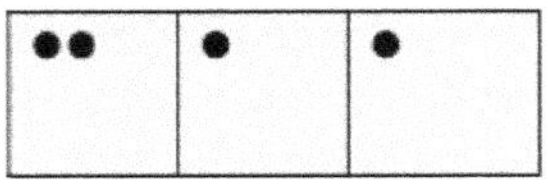

multiply each digit of 23 by each digit of 211, $2 \times 2, 2 \times 1, 2 \times 1, 3 \times 2, 3 \times 1$, and 3×1. Write the results at the intersection of the corresponding row and column. Add the resulting numbers diagonally from left bottom to right top and write them in the last column. In this example, we obtain 4, 853.

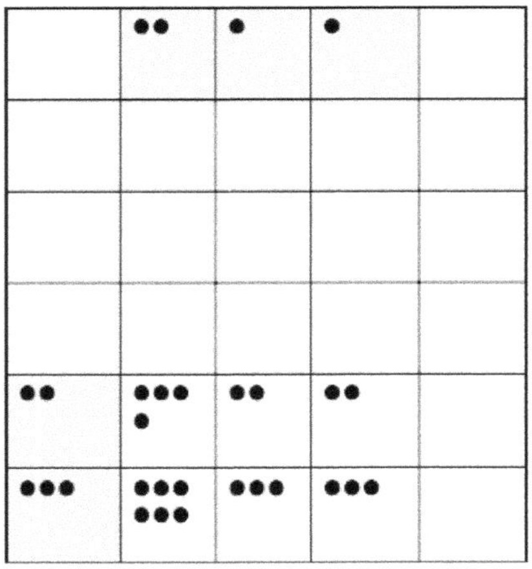 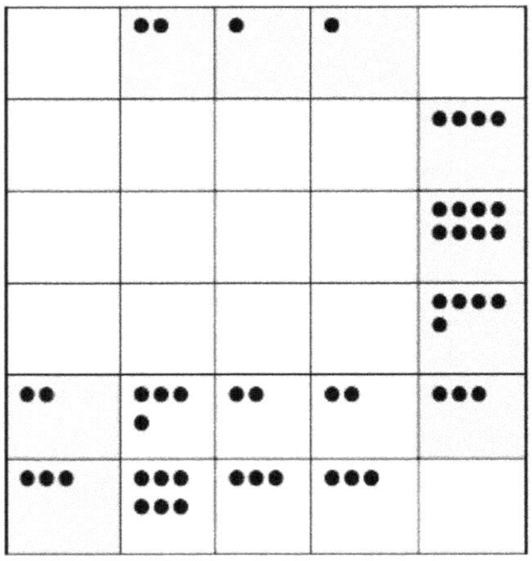

Multipliying With Trades in the Yupana

In the next example, we will need to trade in order to perform the multiplication.

Example 3.3.5

Multiply 43 by 233. After multiplying the digits of 43 by the digits of 233, two of the products are 12, therefore, we have to use the trade rule, see figure below. However, when we add diagonally the resulting numbers, we have to trade three more times to obtain the answer 10, 019.

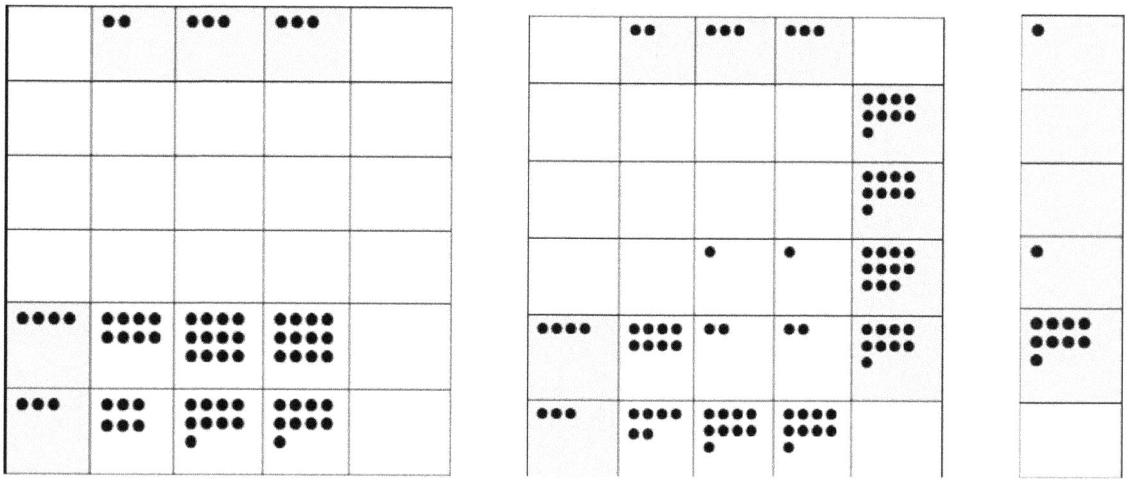

Dividing on the Yupana With No Remainder

As we know, division is "reversing" multiplication. We begin with an example of division without a remainder. Division with remainder will be shown later.

In the division $a \div b = c$, a is called the dividend, b the divisor, and, c the quotient.

Hence, dividend = divisor × quotient.

Example 3.3.6

In this example, we will not have to trade and the remainder will be 0, which is the simplest type of division. Let us divide $4,853 \div 23$. In a previous example, we multiplied 23 by 211 obtaining 4, 853. Therefore, we know that the quotient in this division must be 211.

Step 1: Set-up

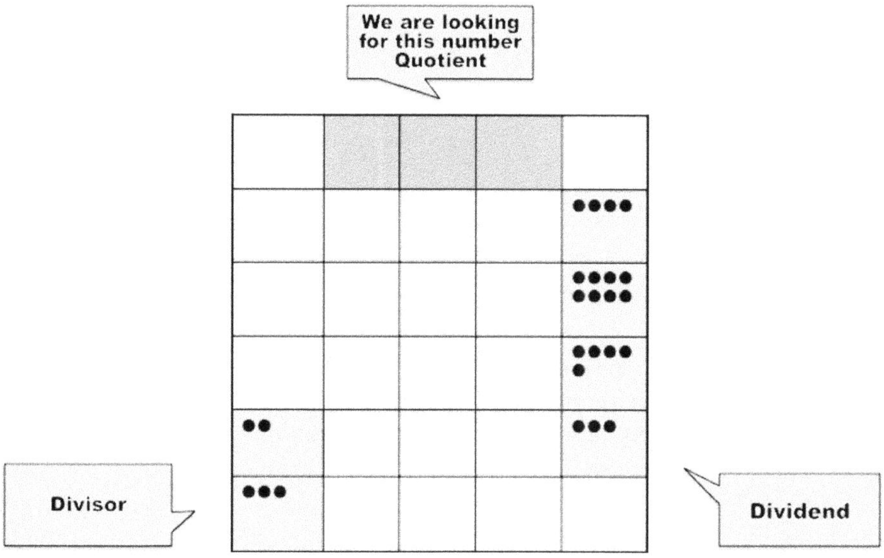

Look at the divisor and try to fit it into the number formed by consecutive digits of the dividend starting from the top: 23 does not go into 4, therefore 23 into 48 goes two times. Place 2 in the top row second column.

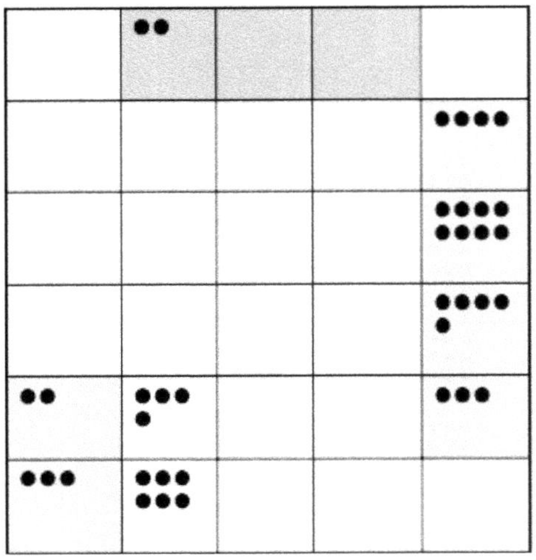

Step 2: Multiply 2 × 23 to get 46, on a side we compute the difference 48 – 46 = 2
Now look at the next digit from the top of the dividend, 5, and form the number 25.

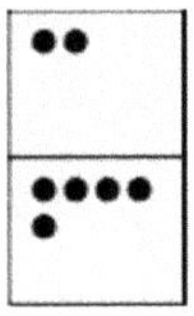

23 goes into 25 one time, we place 1 in the top row, third column, 1 × 23 = 23

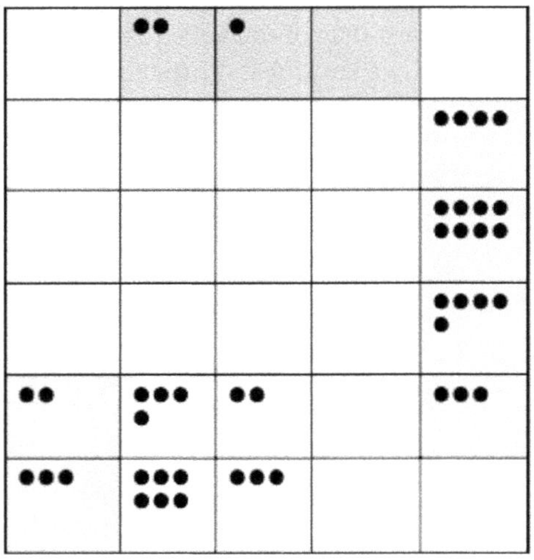

Step 3: Again, on a side, compute the difference 25 – 23 = 2. Take the last digit of the dividend, 3, and form the number 23. Now, 23 goes into 23 one time. Place 1 in the top row, fourth column. Multiply 1 × 23 to get 23.
In this manner, we have recovered exactly the multiplication grid of Example 3.3.4. The top row displays the quotient. Multiplication checks that division is correct.

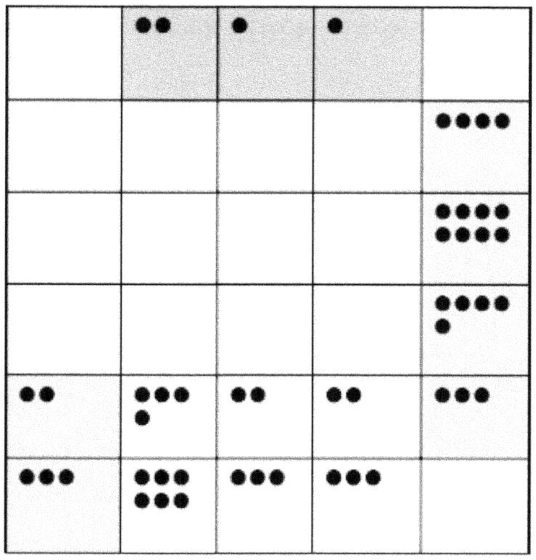

Example 3.3.7

In the example that follows we will have to trade and the remainder will be 0. Divide 1, 378 by 53.

Step 1: Set-up.

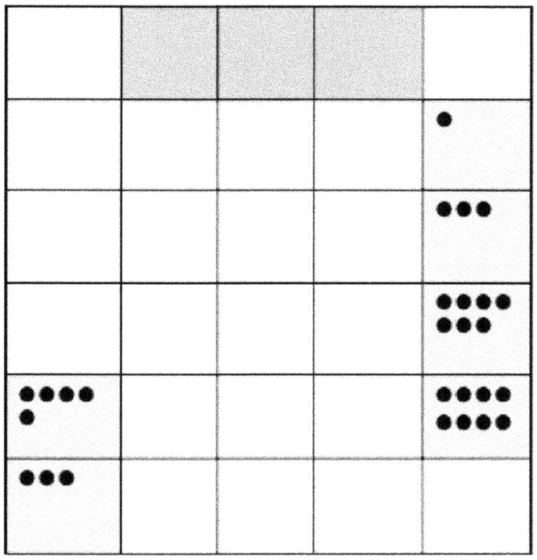

53 goes into 137 two times. We write the number 2 in the quotient.

Step 2: Multiply 2 × 53 to get 106.

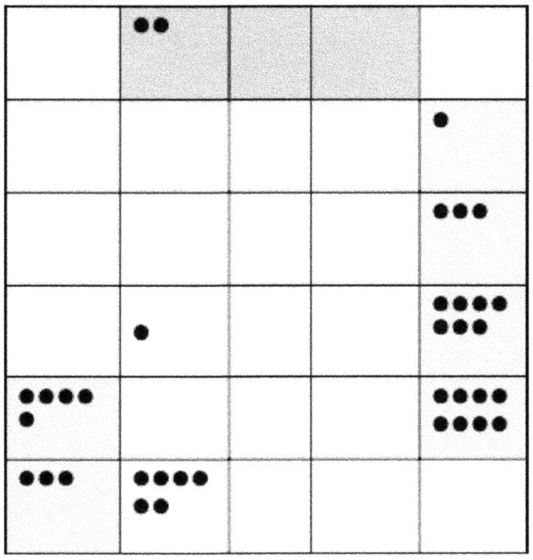

Now subtract 137 – 106 = 31. Get the next (last) digit of the dividend, 8. We have to trade here to write the number 318.

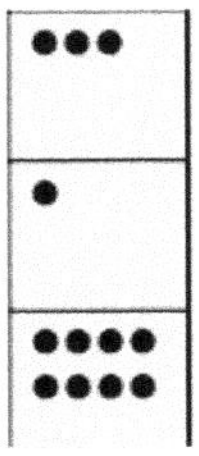

53 goes into 318 six times.

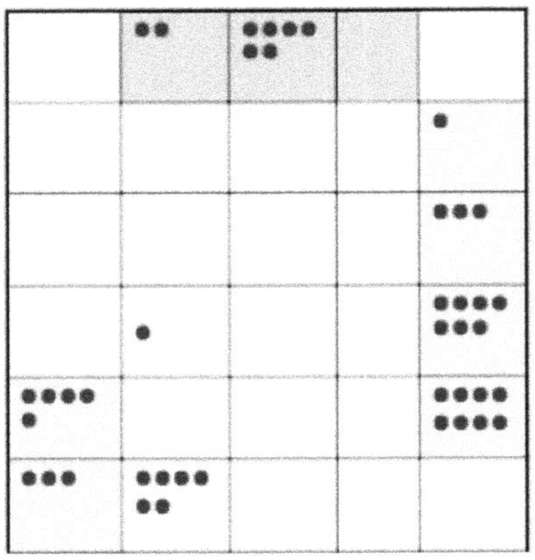

Step 3: Multiply 6 × 53 to get 6 × 5 (tens, which is 3 hundreds) plus 6 × 3 = 18, which means 1 more dot on the row of tens and 8 dots in the row of units (trade).

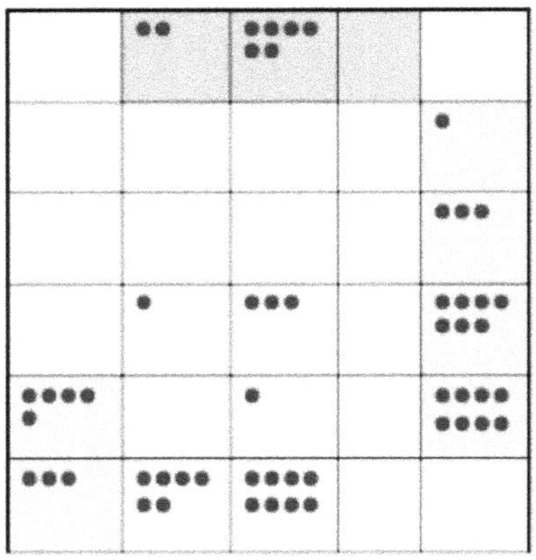

Hence, the quotient is 26.

Dividing in the Yupana With Remainder

In what follows we will use once again the Division Algorithm from Chapter I, Section 1.2.

In general, dividend = divisor × quotient + remainder

Example 3.3.8

In this example, the remainder will be different from 0. Divide 211 by 4. Here, 211 is the dividend and 4 is the divisor, we have to find the quotient and the reminder.

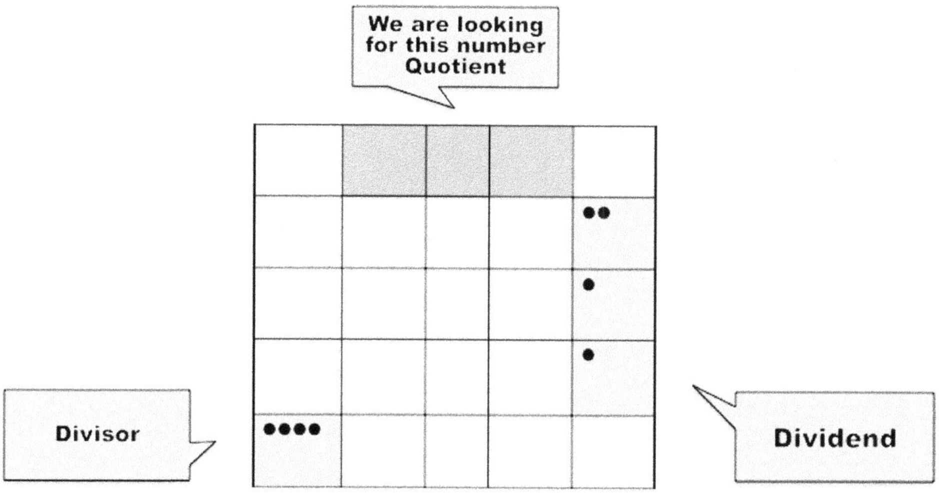

Step 1: 4 goes into 21 five times. We write the number 5 in the quotient and multiply 5 times 4 to get 20 and write it on the grid.

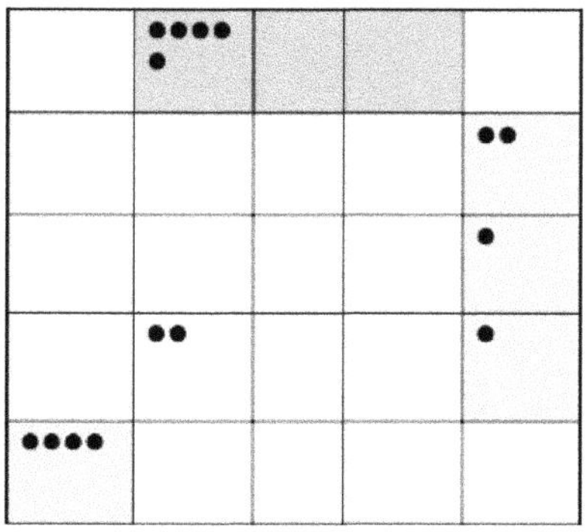

Step 2: Now, we subtract 21 − 20 and obtain 1 and using the units digit 1 in the dividend we form the number 11 and continue as in Step 1. 4 goes into 11 two times and $2 \times 4 = 8$ and we write the numbers 4 in the quotient and 8 in the corresponding cells below 4.

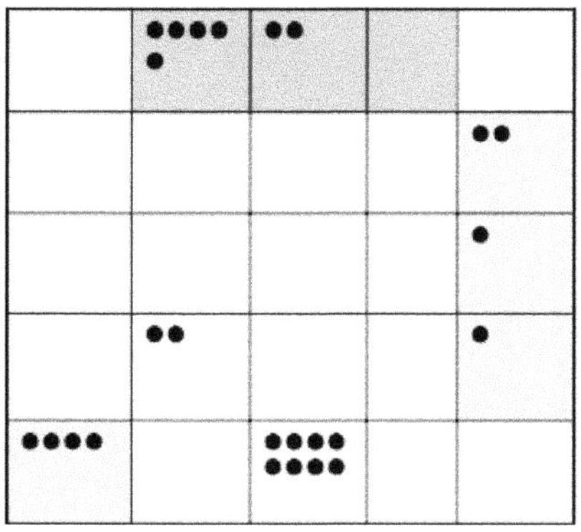

Next, we calculate the difference $11 - 8 = 3$ obtaining the remainder 3. Hence, $211 = 4 \times 52 + 3$.

Example 3.3.9

In this example, we will have to trade and the remainder will be different from 0. Divide 1,328 by 25. We just start as in the previous example.

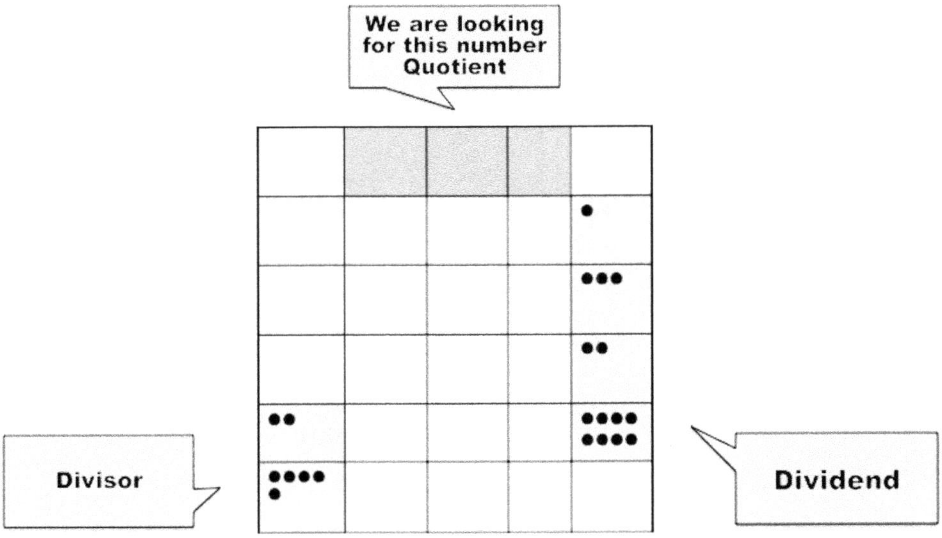

Step 1: 25 goes into 132 five times. We write the number 5 in the quotient.

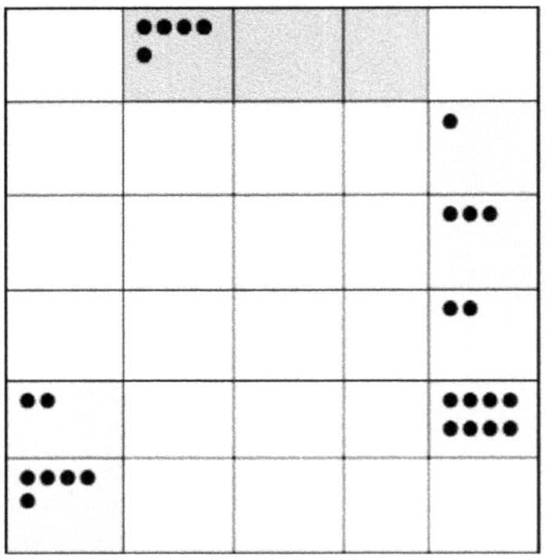

Step 2: Multiply 5 × 25 to get 125 and write the number below the 5 we have in the divisor.

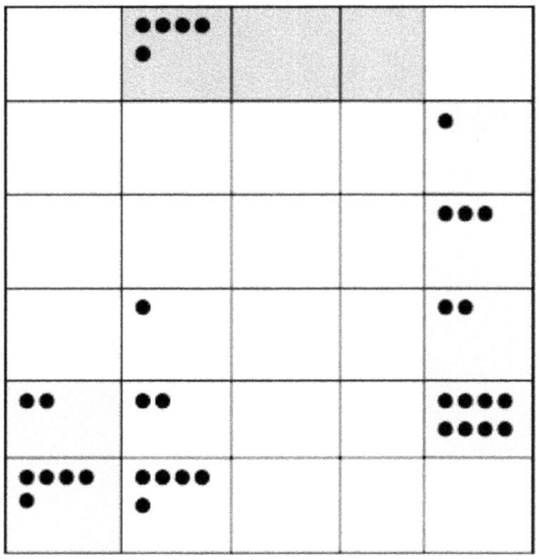

Now we have to subtract 132 – 125 = 7, and use this number and the units digit in the dividend to form 78, and continue the process as in Step 1.

The divisor 25 goes into 78 three times and we write the numbers 3 in the divisor and the product $3 \times 25 = 75$ in the same column below 3.

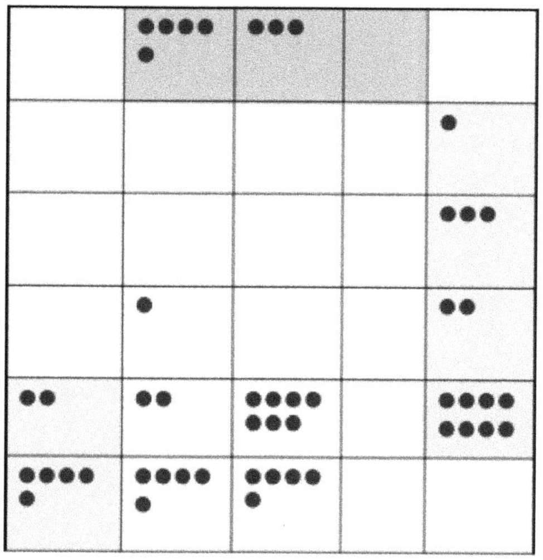

Now, $78 - 75 = 3$ and therefore, the remainder in the division is 3.
Hence, $1,328 = 25 \times 53 + 3$.

EXERCISES

1. Write the following numbers in Quechua.

   a. 23 **b.** 124 **c.** 1,704
   d. 2,222 **e.** 6,600 **f.** 21,345
   g. 100,919 **h.** 452,000 **i.** 90,000,000

2. Write the following numbers in Hindu-Arabic:

   a. iskay chunka ujniyuq
   b. kinsa pachaq chunka pusaqniyuq
   c. tawa pachaq suqta chunka jisqoniyuq
   d. waranqa pachaq chunka pusaqniyuq
   e. kinsa hunu pusaq waranqa jisqon pachaqnin
   f. tawa waranqa tawa pachaq tawa chunka tawayuq
   g. suqta hunu pachaq jisqon chunkan
   h. kinsa chunka kinsayuk waranqa kinsa pachaq kinsa chunka kinsayuq
   i. iskay waranqa kinsa pachaq suqta chunka pusaqniyuq

3. Draw a quipu representing all the numbers obtained in exercise 2.
4. Draw a quipu representing the dates the last 6 Inca kings reigned:

   Viracocha Inca

   Pachacuti Inca Yupanqui

   Tupac Inca Yupanqui

   Wayna Capac

   Huascar

   Atahualpa
5. Use a grid to write the beginning of Viracocha Inca's reign, 1430, using Wassen's method.

6. Utilize a grid to write the year the ruler Atahualpa was executed, 1533, using the Gheverghese method. Below is one of the many answers. Find another representation of that year.

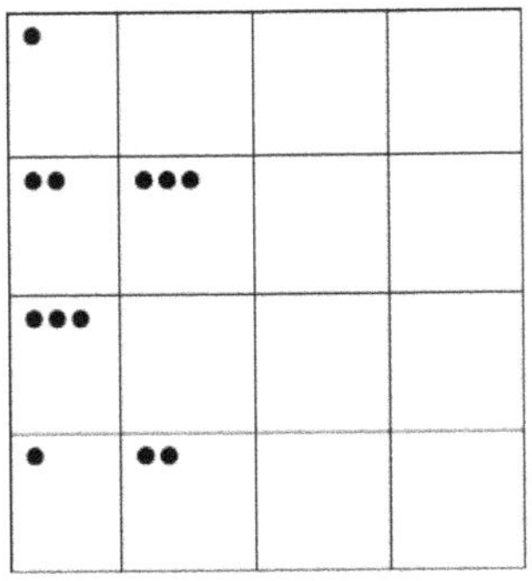

7. Find the numbers represented on the grids using the Wassen and Gheverghese methods.

a.

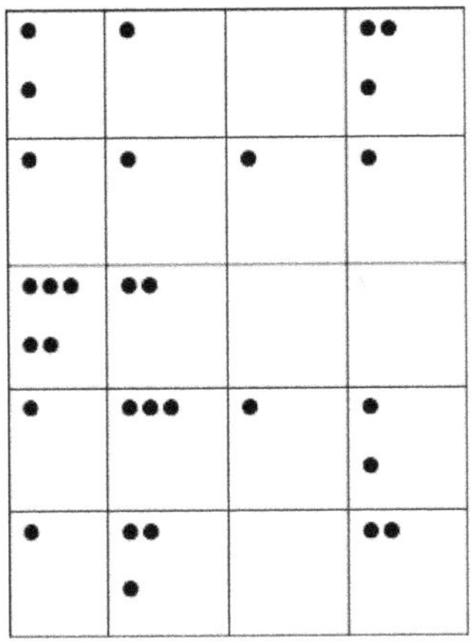

b.

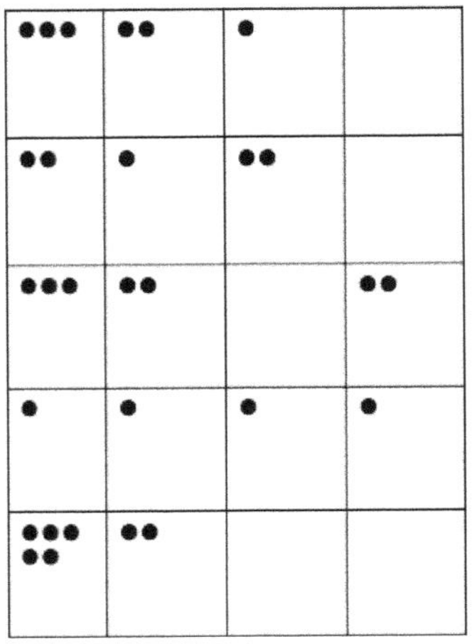

c.

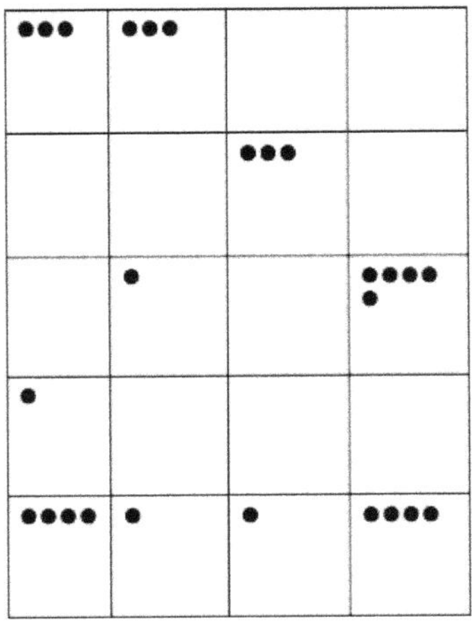

8. Draw a grid to add the following numbers.

a. 45, 678 and 123
b. 31, 425 and 7, 423
c. 68, 670 and 9, 876
d. 10, 299 and 33, 490

9. Draw a grid to subtract the following numbers.

a. 777 from 3, 121
b. 1, 213 from 5, 678
c. 937 from 1, 952
d. 9, 999 from 10, 789

10. Draw the grids needed to multiply the following numbers.

a. 27 by 311
b. 45 by 213
c. 71 by 119
d. 92 by 545

11. Draw the grids needed to divide the following numbers.

a. 2, 268 by 84
b. 2, 442 by 22
c. 1, 234 by 15
d. 1, 253 by 33

12. Write the answers obtained in problem 10 using Quechua.
13. Write the answers obtained in problem 11 using Quechua.
14. Draw a quipu representing the following information from your class:

a. Number of students
b. Gender
c. Age
d. Major

15. Draw a quipu recording the information from the cash register in the figure below.

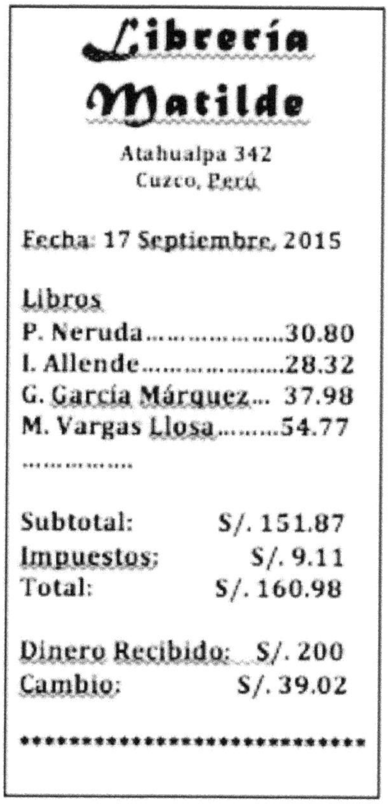
Librería
Matilde
Atahualpa 342
Cuzco, Perú

Fecha: 17 Septiembre, 2015

Libros
P. Neruda....................30.80
I. Allende.....................28.32
G. García Márquez... 37.98
M. Vargas Llosa.........54.77
....

Subtotal:	S/. 151.87
Impuestos:	S/. 9.11
Total:	S/. 160.98
Dinero Recibido:	S/. 200
Cambio:	S/. 39.02

****

16. Write your own detailed interpretation of the following quipus.

a.
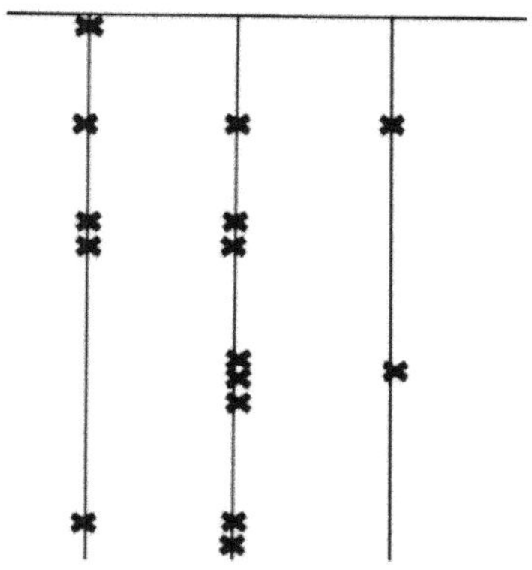
b.
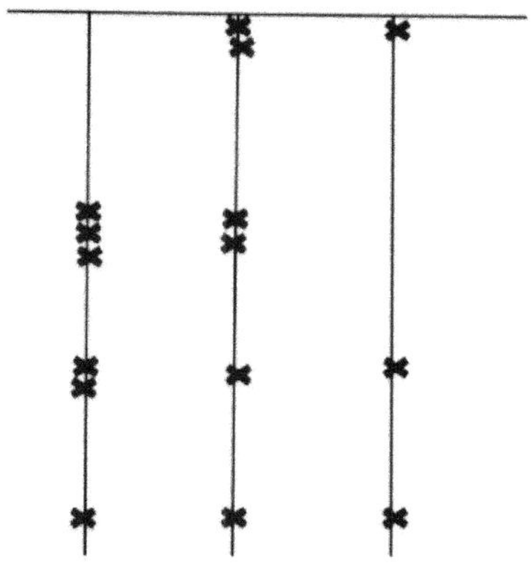
c.
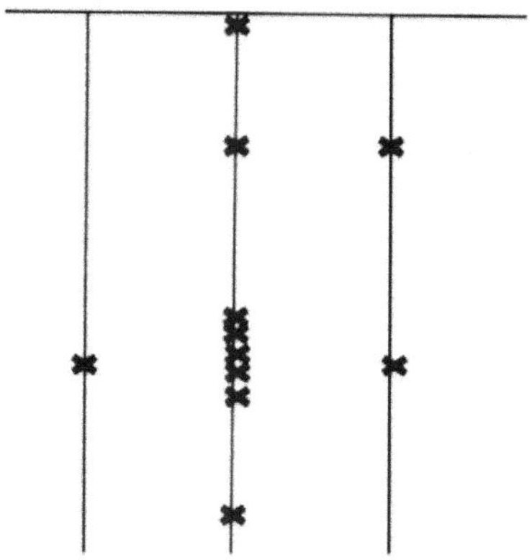

REFERENCES

Acosta, de Jose. 1940. *Historia Natural y Moral de las Indias*. Mexico. Edicion.

Barnhart, Edwin. 2012. *Lost Worls of South America [Videorecording]* Chantilly, VA: The Teaching Co.

Bauer, Brian and David Dearborn. 1995. *Astronomy and Empire in the Ancient Andes*. Austin: University of Texas Press.

Calderon, Hector. (1966). *La Ciencia Matemática de los Mayas*. México, D.F.: Editorial Orion.

Catepillán, Ximena and Waclaw Szymanski. 2012. "Counting and Arithmetic of the Inca. *Revista Latinoamericana de Etnomatemática* 5(3): 47–65.

Gheverghese, George. 1990. *The Crest of the Peacock*. London: Penguin Books.

González Holguín, Diego. 1989[1608] .*Vocabvlario de la Lengva General de todo Perv llamada lengua Qquichua, o del Inca.* Lima, Perú: Editorial de la Universidad Nacio-nal Mayor de San Marcos.

Guaman Poma de Ayala, Felipe. 1936. *Nueva Coronica y Buen Gobierno.* Paris: Institut d'Ethnologie, Universite de Paris.

Inca Garcilaso de la Vega. 2004. *Comentarios Reales de los Incas.* Peru: A.F.A. Editores Importadores S.A., Edicion.

Radicati Di Primeglio, Carlos. 1973. *El Sistema Contable de los Incas.* Mexico: Libreria Studium.

Urton, Gary. 1997. *The Social Life of Numbers: A Quechua Ontology of Numbers and Philosophy of Arithmetic.* Austin: University of Texas Press.

Wassen, Henry. 1931. "The Ancient Peruvian Abacus." In *Comparative Ethnographical Studies,* edited by Erland Nordenskiold. Vol. 9. Goteborg: AMS Press.

ONLINE RESOURCES

Museo Chileno de Arte Precolombino. http://www.precolombino.cl/

Chapter 4

The Arithmetic of the Maya

"ah bolon hobon"

one who knows many trades

> *The nations of Yucatan . . . had a high priest whom they called Ah Kin (Lord of the Days) . . . In him was the key of their learning and it was to these matters that they dedicated themselves mostly . . . They provided priests for the towns when they were needed, examining them in the sciences and ceremonies, and committed to them the duties of their office, and the good example to people and provided them with the books and sent them forth. And they employed themselves in the duties of the temples and in teaching their sciences as well as in writing books about them.* (Tozzer 1941, pp. 27–28).

The civilization of the Maya, which began ca. 1200 BC, had a long history divided into three periods: The Pre-Classic period ca. 1200 BC–200 AD, the Classic period 200 AD–900 AD, and the Post-Classic period 900 AD–1519 AD (Stuart, 2011). This civilization extended from what it is now Belize, Central and Southern Mexico, Guatemala, El Salvador and parts of Honduras. About 30 Mayan languages are still spoken in the abovementioned regions, with a Maya descendants' population of about five million, two million of which follow Maya traditions. The general conventions used lately by anthropologists, linguists, and archaeologists states that the words Maya and Mayan are not synonymous, the word Maya is a general ethnic term and adjective, for example, Maya number, whereas the word Mayan is reserved for the language group, for example, Mayan dialect (Stuart, 2011). The Late Classic period witnessed its peak population, and the political, social, artistic, and intellectual highpoints varied from region to region. The Maya developed a glyph writing system, a number system, and they mastered several areas of science, art, and architecture. Perhaps the most significant developed area was astronomy, which was important for agriculture and everyday life. Based on astronomical observations, the Maya created an elaborate system of calendars. The most studied calendars are the Tzolkin, the Haab, and the Long Count, the well known Round Calendar is a combination of the first two. They also developed at least three more calendrical cycles: the Venus cycle, the Lunar calendar, and an 819-day count ritualistic cycle. The next chapter is devoted to the Maya calendars and some of the conversions among Tzolkin, Haab, and Long Count dates. The Mathematics the Maya used for their astronomical calculations utilized a vigesimal number system, which is described in this chapter together with the basic arithmetical operations. For their calendrical representations, they used an "almost" vigesimal system, called chronological system by Sylvanus G. Morley in 1946.

The map in Figure 4.1 highlights the better-known Maya sites.

4.1 The Maya Counting System

Digits are sign-symbols used to write numbers. In our present writing of numbers we use the Hindu-Arabic system. It has two basic characteristics: it is decimal and positional. The base of this system is 10, for which the word in Latin is *decem*. Hence, the name of the system: *decimal*. The decimal system requires 10 symbols—digits which are: 0, 1, 2, 3, 4, 5, 6, 7, 8, 9. The system is positional, which means that with these digits properly positioned any number can be written. Consecutive positions are ordered from right to left: units, tens, hundredths, and so on. For example, the

Figure 4.1: Map of the Maya world

number 123 has 3 units, 2 tens, and 1 hundred, which means: $123 = 1 \times 10^2 + 2 \times 10 + 3 \times 1$. Notice that the digits are positioned horizontally.

The Maya, like other peoples of the pre-Columbian Mesoamerica, developed a counting system with two basic characteristics: it is vigesimal and positional. The base of the system is 20, for which the word in Latin is *viginti*. Hence, the name of the system: *vigesimal*. The vigesimal system requires 20 digits: 0, 1, ..., 19, which are presented in Figure 4.2. Perhaps the reason for which some of the pre-Hispanic cultures chose the vigesimal system is that we can not only count with our fingers, but also with our toes. The Maya notation for the digits is very efficient because there are only three basic symbols: a dot for one, a bar for five, probably sticks and pebbles or beans, and a symbol for zero. As we will see later, the Maya numerical system was not entirely vigesimal when used for calendrical purposes.

The Maya or Mesoamerican zero dates back to 100–200 BC, while the Babylonian or Mesopotamian dates back to 300–400 BC. The ancient Egyptians, Greeks, Romans, and Chinese had not developed zero at the time. The creation of zero was the key to develop their positional counting system. Without this important discovery, there was not much hope for the development of the mathematics and therefore engineering and the sciences. The Maya devised several symbols for zero, a special group of zeros were used in the Maya writings, the codices, and another set for the inscriptions, see Figure 4.2 with the Maya digits. The most common symbol for zero, which will be used in our computations, was , which is believed to represent a sea shell in Figures 4.2 and 4.3.

The way the Maya positioned the digits is different from the way of positioning the digits in our Hindu-Arabic only in a design type aspect, not intrinsically. While in our system we position the digits horizontally, the Maya positioned their digits vertically. Some exceptions are known: when digits were used to represent Maya dates in the inscriptions the direction of reading sometimes was from top to bottom. To avoid confusion in this latter case, the Maya attached to the digits a period glyph which indicated the corresponding multiplier.

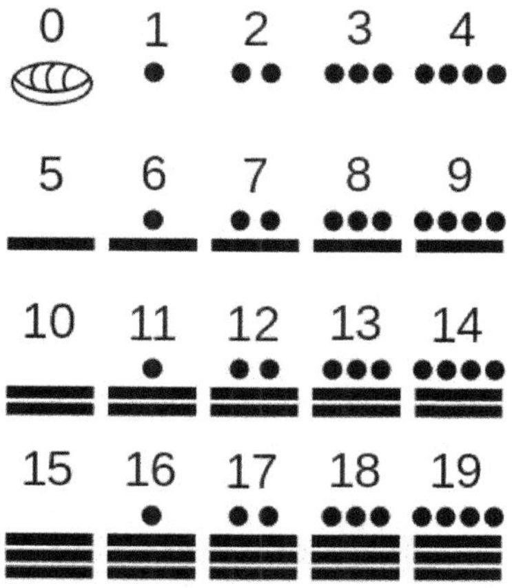

Courtesy of Nicholas Heil

Figure 4.2: The Maya digits, drawing by Nicholas Heil, 2015

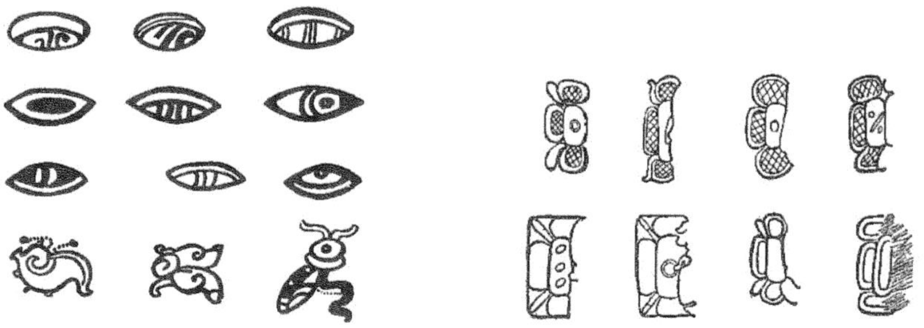

From *An Introduction to the Study of the Maya Hieroglyphs* by Sylavsanus G. Morley. Copyright © 1975 by Dover Publications, Inc. Reprinted by permission.

Figure 4.3: Symbols for zero in the codices left, and for the inscriptions at right

In what follows all the numbers will be read from bottom to top, as in the codices, the Maya books.

For example, to represent the number 28 they used a dot—the digit one (above) and the symbol—digit for eight (below). Here on the top place value we have $1 \times 20^1 = 20$, and at the bottom $8 \times 20^0 = 8$, therefore the number is $28 = 1 \times 20^1 + 8 \times 20^0$.

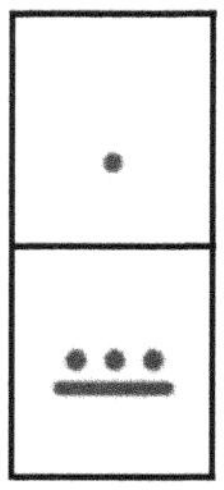

As another example, the Maya representation of 380 is $380 = 19 \times 20^1 + 0 \times 20^0$. That is,

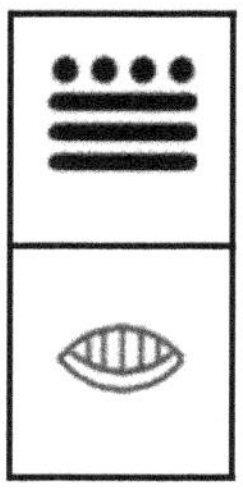

To represent 563, we write $536 = 1 \times 20^2 + 8 \times 20^1 + 3 \times 20^0$. Thus, we need the symbols for 1, 8, and 3.

In the chronological system, the difference is at the third digit, which is multiplied by 18 × 20 as opposed to 20 × 20, the fourth and higher digits continued to be multiplied by powers of 20. Table 4.1 illustrates the two systems.

The chronological system was used to represent calendrical information in the codices—the Maya books, the Books of Chilam Balam—compilations from older and now lost manuscripts, and in their inscriptions. For the latter they used sometimes a similar system in which the direction of reading, bottom to top, was reversed (Morley, 1975).

The Maya also used head profiles representing different Gods to exemplify the numbers in which the essential differences were in the forehead, headdresses, and facial characteristics. See Figure 4.4.

Notice in Figure 4.4 that the numbers 2, 6, 7, 11, 12, 16, and 17, have the spaces between the dots filled with dot-like ornaments. Apparently this was done by the artists for aesthetic reasons.

Moreover, besides the dots, bars, the symbol for zero, and the head profiles, they used symbolic variants for the digits 1–19, and also for 20. Figure 4.5 depicts the symbolic variants for some of the digits.

Some Mayanists claim that the number system is a quinary-vigesimal system of counting, that is, using a rule of five probably based on finger counting. In Latin, the word for 5 is "*quinque*," hence the name. The Roman number system is an example.

Table 4.1: The vigesimal and chronological systems

Units	Vigesimal System		Units	Chronological System	
$20^3 = 8{,}000$	•	═	$18 \times 20^2 = 7{,}200$	•	═
$20^2 = 400$		•	$18 \times 20 = 360$		•
$20^1 = 20$		•••	$20^1 = 20$		•••
$20^0 = 1$	• —	—	$20^0 = 1$	• —	—
	8,006	80,465		7,206	72,425

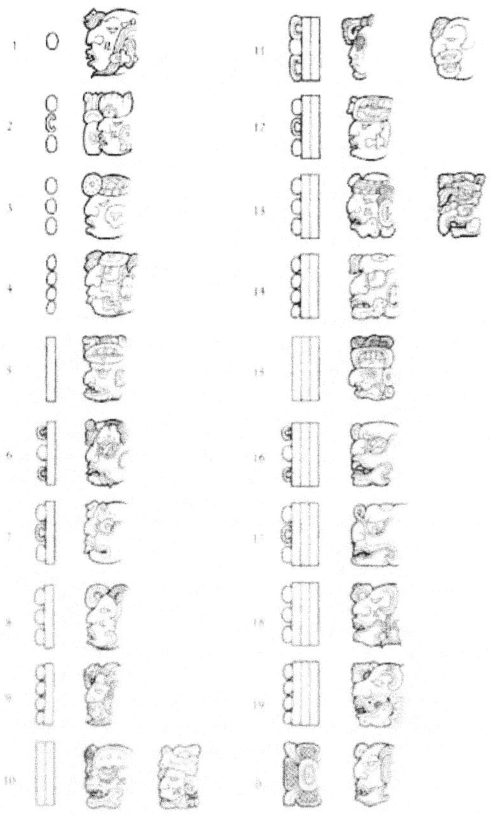

Illustrations by Mark Van Stone from *Reading the Maya Glyphs* by Michael D. Coe, Thames & Hudson, London and New York.

Figure 4.4: Head variants with corresponding numbers (Coe, 2001)

Maya Hieroglyphic Writing by J. Eric S. Thompson.

Figure 4.5: Symbolic variants for the numbers 9, 18, 16, 19, 1, 3, and 4 from left to right (Thompson, 1971)

The Spanish missionary Diego de Landa (1524–1579), infamously known for the burning of thousands of Maya books (codices) and religious images, in his attempt to convert the Maya people to Christianity. Only four codices survived. The missionary, and later elected Bishop of Yucatan, who arrived in America in 1549, wrote:

Que su contar es de 5 en 5 hasta 20, y de 20 en 20 hasta 100, y de 100 en 100 hasta 400, y de 400 en 400 hasta 8 mil; y de esta cuenta se servian mucho para la contratacion del cacao. Tienen otras cuentas muy largas y que las extienden ad infinitum contando 8 mil 20 veces, que son 160 mil, y tornando a 20, duplican estas 160 mil, y despues de irlo asi duplicando hasta que hacen un incontable numero, cuentan en el suelo o cosa llana. (De Landa, 1566)

Their count is by fives up to twenty, and by twenties up to one hundred and by hundreds up to four hundred, and by four hundreds to eight thousand; and they used this method of counting very often in the cacao trading. They have other very long counts and they extend them in infinitum, counting the number 8000 twenty times, which makes 160,000; then again this 160,000 by twenty, and so on multiplying by 20, until they reach a number which cannot be counted. They make their counts on the ground or something smooth. (Tozzer, 1941)

The number names in one of the various Mayan languages, the Yucatec, are presented in Table 4.2 (Beltran de Santa Rosa 1859, Ifrah 2000). Notice that the spoken numbers in Yucatec made use of a vigesimal system for counting, in fact, all of the Mayan dialects made use of a vigesimal system for reckoning.

Table 4.2: Maya numbers in Yucatec

	Yucatec		Yucatec		Yucatec		Yucatec
1	hun	**14**	can lahun	**36**	uac lahun tu kal	**87**	uuc tu y ho kal
2	ca	**15**	hol hun	**40**	ca kal	**90**	lahun tu y ho kal
3	ox	**16**	uac lahun	**41**	hun tu y ox kal	**99**	bolon lahun tu y ho kal
4	can	**20**	hun kal	**42**	ca tu y ox kal	**100**	ho kal
5	ho	**21**	hun tu kal	**54**	can lahun tu y ox kal	**111**	buluc tu y uac kal
6	uac	**22**	ca tu kal	**59**	bolon lahun tu y ox kal	**120**	uac kal
7	uuc	**23**	ox tu kal	**60**	ox cal	**133**	ox lahun tu y uuc kal
8	uaxac	**30**	lahun ca kal	**61**	hun tu y can kal	**140**	uuc kal
9	bolon	**31**	buluc tu kal	**64**	can tu y can kal	**200**	lahun kal
10	lahun	**32**	lahca tu kal	**70**	lahun tu y can kal	**400**	hun bak
11	buluc	**33**	ox lahun tu kal	**79**	bolon lahun tu y can kal	**800**	ca bak
12	lahca	**34**	can lahun tu kal	**80**	can kal	**8,000**	hun pic
13	ox lahun	**35**	holhu ca kal	**81**	hun tu y ho kal	**16,000**	hun calab

Numbers 1–11 have names listed in Table 4.2. Numbers 12–19 are formed by adding the numbers 2–9 to 10 (*lahun*). Number 12, *lahca* is *lahunca* abbreviated. In several Mayan dialects, the word *kal* for 20 means "*one score*" and in others "*one man*." The numbers 21–39 are formed by adding the unit name (1–19) to 20 (*kal*) using the word "*tu*" to denote the addition (after). For example: 26 = *uac tu kal* = *six added to twenty* = *six after twenty*.

The numbers 31–39 are represented as 11 after 20, 12 after 20, ...,19 after 20, with the exception of 35 = *holhu ca kal* = *fifteen two twenty* wich apparently means *fifteen* in the *second twentieth*. Numbers 41–59 are all written as 1 in the third twentieth, 2 in the third twentieth, ... 19 in the third twentieth, by adding the suffix "*tu y*" before "third twentieth."

Notice that for the second powers of 20, that is, numbers 20–399, the suffix *kal* is attached to the numeral, for example:

40 = *two twenties* = *ca kal*,

100 = *five twenties* = *ho kal*, and

30 = *ten lacking to two twenties* = *lahu ca kal*.

The suffix *bak* is attached to the third powers of 20 that is, numbers 400–7999, and *pic* and *calab* to the fourth and fifth powers of 20, respectively. As a rule, added digits are prefixed in Yucatec, for example: *can lahun* = *four added to ten* = 14, but in other Mayan languages the reverse rule applies. (Thompson, 1971).

4.2 Arithmetic Operations

To perform the mathematical operations, we will represent the numbers only in vigesimal form, that is, using powers of 20.

To add numbers in vigesimal form, we just add the corresponding places and remember to trade a digit from a lower to a higher place when the number is greater than 19. This is similar to our Hindu-Arabic addition in which case when the number is greater than 9 we carry a digit from a right to a left place.

Example 4.2.1

To add the numbers 66 and 28, we do not need to trade since none of the place values is greater than 19. Let us represent the numbers in powers of 20. $66 = 3 \times 20^1 + 6 \times 20^0$, $28 = 1 \times 20^1 + 8 \times 20^0$.

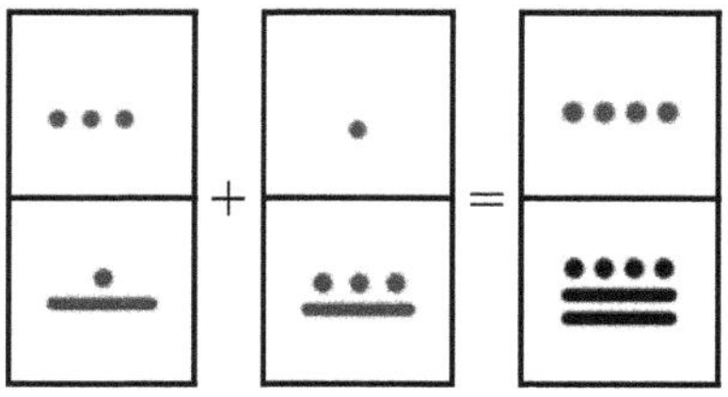

The answer is $4 \times 20^1 + 14 \times 20^0 = 94$.

A remark on trading: If when perfoming an operation using Maya digits we obtain a number 20 or greater in one cell, we will need to trade because such number is not a Maya digit (similar to the trading we do with Hindu-Arabic digits). In the next example, we will need to trade because in one of the cells we get the number 22 which is not a Maya digit.

Example 4.2.2

To add the numbers 2,003 and 7,049 we will need to trade to obtain 9,052. First, we represent the numbers in powers of 20.

$2{,}003 = 5 \times 20^2 + 0 \times 20^1 + 3 \times 20^0 = 5 \times 400 + 3 \times 1$ and $7{,}049 = 17 \times 20^2 + 12 \times 20^1 + 9 \times 20^0 = 17 \times 400 + 12 \times 20 + 9$. Notice that we get 22 in the top cell, and since this is not a Maya digit we need to trade after writing the number as $20 + 2$.

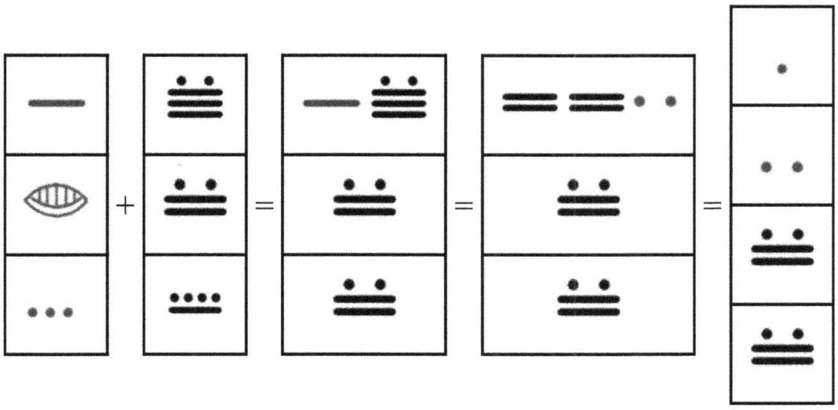

The answer is $1 \times 20^3 + 2 \times 20^2 + 12 \times 20^1 + 12 \times 20^0 = 9{,}052$.

The subtraction can be performed in a similar fashion.

Example 4.2.3

Let us subtract 65 from 202. The corresponding Maya numbers are 202 = 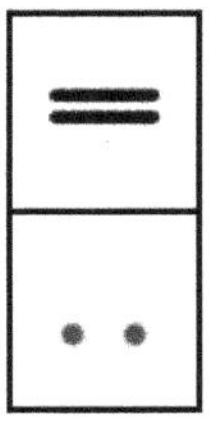 and 65 = 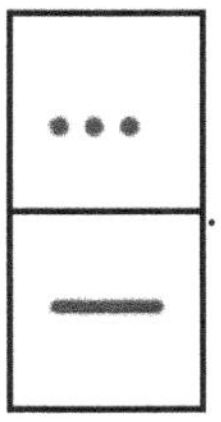. To compute the difference 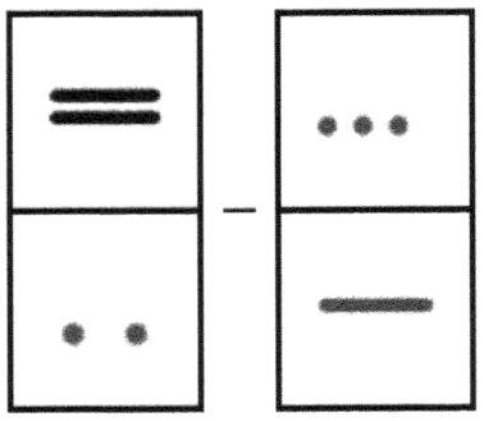 we have to trade since we cannot subtract 5 from 2. Therefore, will rewrite the number

202 as 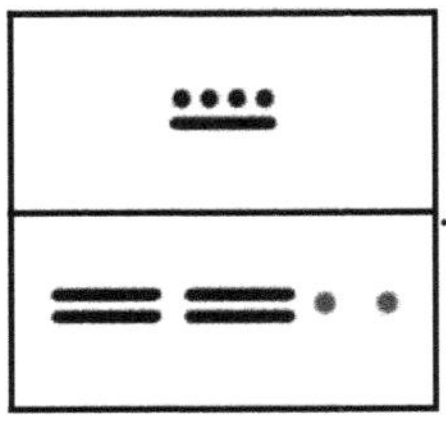.

Hence,  which is 137.

Multiplication

In the Dresden codex, we can find the 4, 5, 6, and 7 × 13 multiplication tables. Tables of multiples of 78, 91, and 364 can also be found with several other tables in connection with almanacs describing mainly weather forecasting and crop predictions. The Maya must have used successive sums to perform the basic multiplications, that is, the product of any quantity by a single digit.

Figure 4.6 shows the middle section of page 44 from the Dresden Codex which contains a table of multiples of 78 using the chronological/calendrical system. In the lower half of the middle section, the list of Maya numbers are from left to right: 10 × 78 (here the bottom digit is zero in red), and in black: 5 × 78,4 × 78,3 × 78,2 × 78, and 1 × 78. In the upper half of this middle section, are the 6,7, 8, and 9 multiples of 78, together with the products 50 × 260 and 60 × 260. The animal like figure on the right is God K, the God of the storm (Multuldzuc), also known as "the Mars beast," this ubiquitous God is characterized by its ornamented nose. (Villacorta and Villacorta, 1930).

The table of multiples of 78 may have its origins on the 780-day synodic period of Mars that they calculated. (The actual period is 779.94 days.) Multiples of 78 can also be found on page 59 of the Dresden Codex following a table of lunar eclipses.

To multiply Maya numbers, we will follow a method developed by Hector M. Calderon (Calderon, 1966). For factors with more than one digit, Calderon suggested they did their computations using a grid, corn kernels (zea mays), and/or beans (buul). Corn and beans were the major Maya food sources. There is some supporting evidence since archaeologists have found many grid patterns scratched into stucco floors, moreover, modern Maya Shamans in the Guatemalan Highlands called "Day Keepers" nowadays use a grid and beans to count through the days of the 260 day sacred calendar. There is no evidence of the specific method used by the Maya but we know they used multiplication and division to handle their calendrical computations. As in addition and subtraction, we will also have to pay attention to trading digits when needed.

Let us multiply a few numbers using the vigesimal Maya number system and Calderon's method.

Example 4.2.4

Multiply 66 by 22. First, we write the numbers using Maya symbols: 66 = 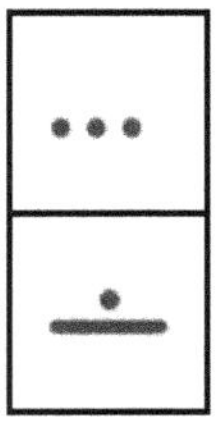 and 22 = 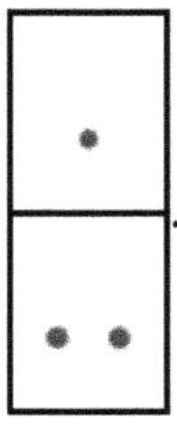. Then, we locate the numbers outside the table below.

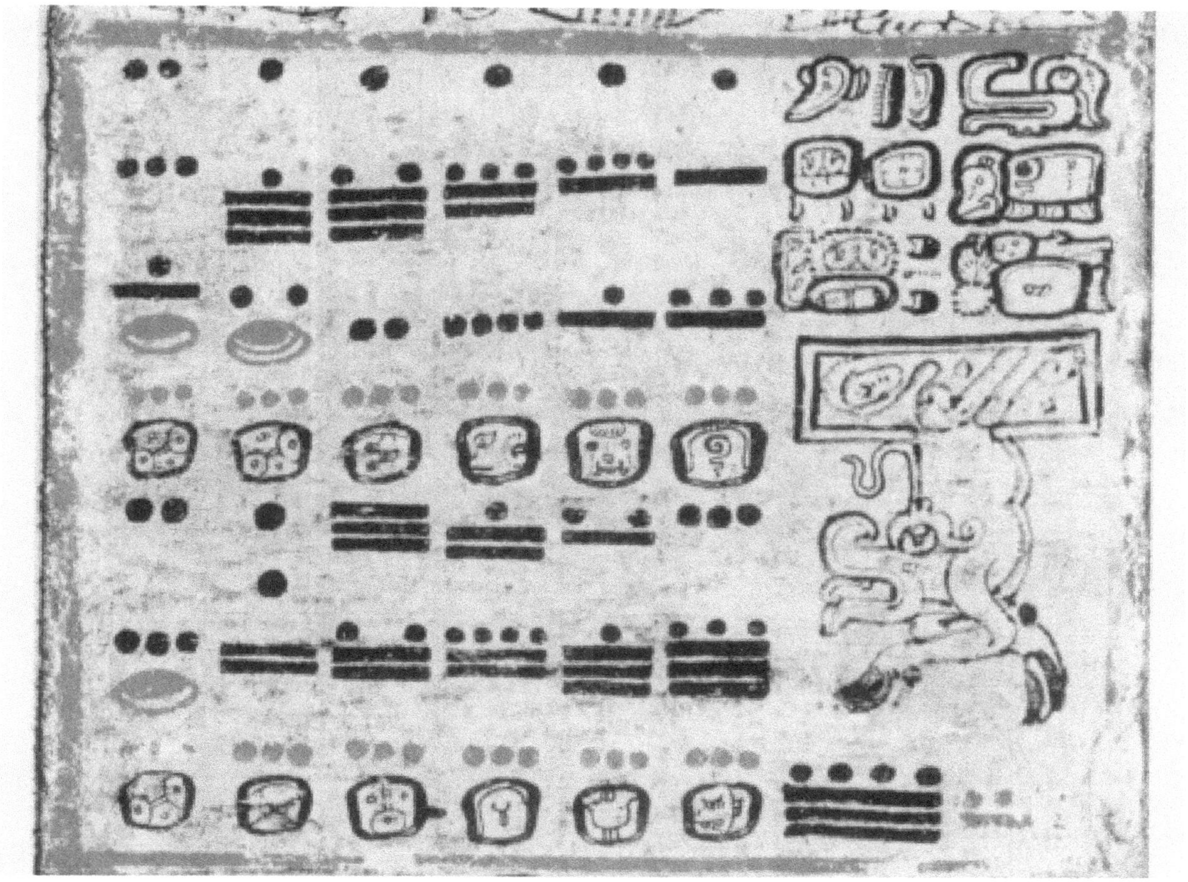

Figure 4.6: Dresden Codex

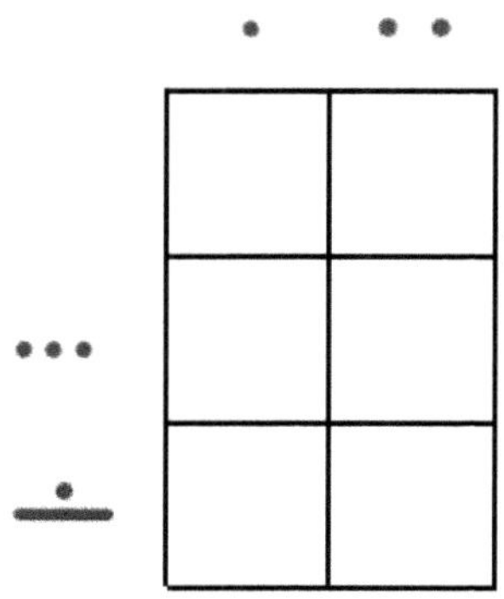

Next, we multiply the corresponding cells 3×1, 3×2, 6×1, and 6×2.

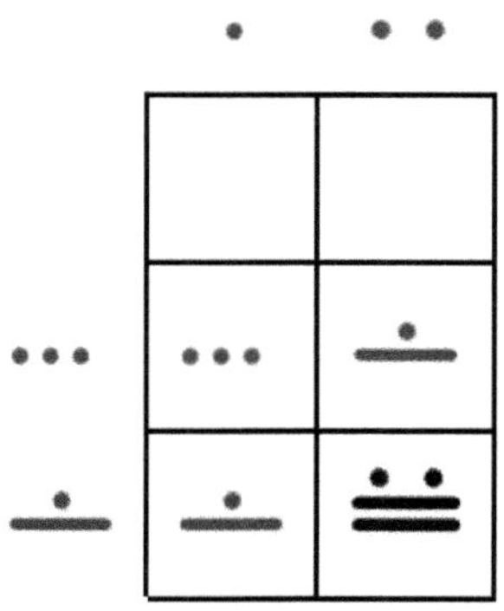

Next, we add the numbers in each one of the diagonals, from lower left to upper right, inside the table obtaining , 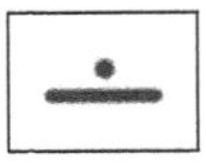 + 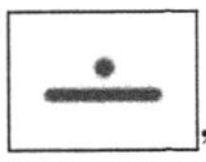, and . The answer is then , which is the number $3 \times 20^2 + 12 \times 20^1 + 12 \times 20^0 = 1{,}452$.

The final table looks as follows.

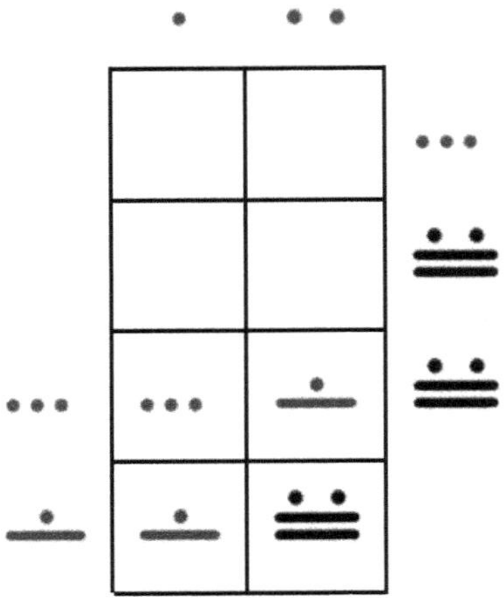

In the following example, we will need to trade in order to perform the multiplication.

Example 4.2.5
Multiply 62 by 88. Here $62 = 3 \times 20 + 2$, and $88 = 4 \times 20 + 8$. And in Maya symbols we have

62 = 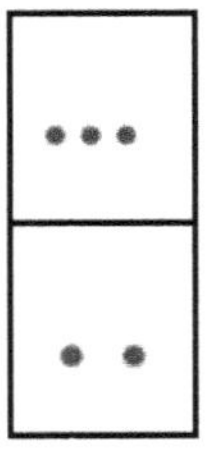 and 88 = 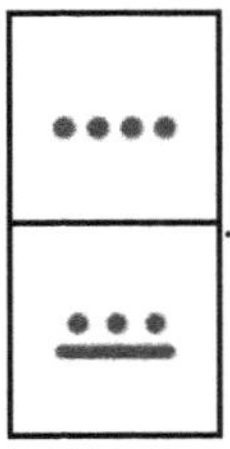. Now, we arrange the numbers around the table and complete the corresponding products as follows.

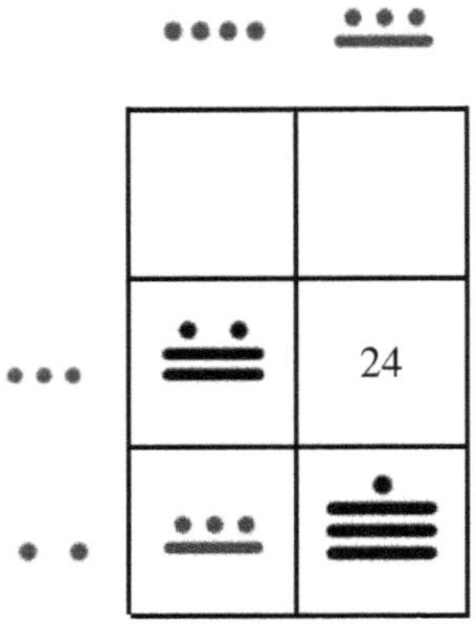

And since we obtained 24 in one of the cells, we must trade and rewrite 24 as

20 + 4, which in Maya symbols is 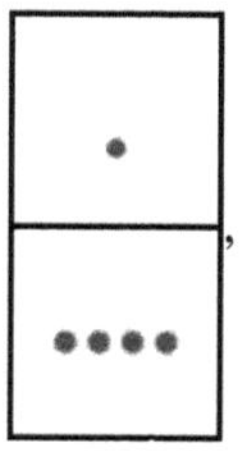, (cf. remark on trading). The completed table looks now as follows.

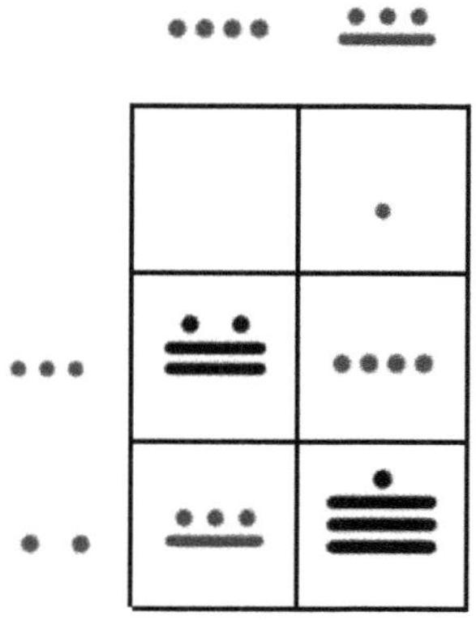 Now, we add the corresponding numbers in each of the diagonals  + 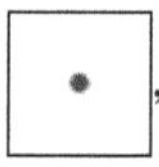,  + 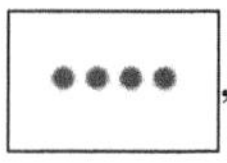, and, , obtaining the Maya number .

This is how the final table looks:

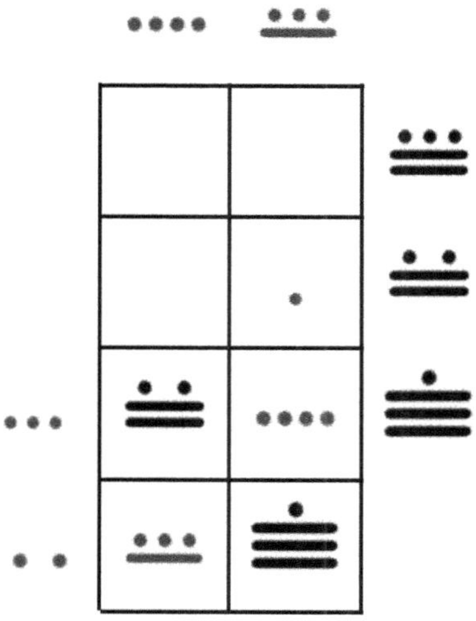

The Hindu-Arabic number is 5,456, the product of 62 and 88.

Division

Division is an arithmetic operation no less important than addition, subtraction, or multiplication. It is common in everyday life when it is necessary to separate a larger quantity of items into smaller groups. In practice, we deal mostly with whole numbers: 0, 1, 2, 3, 4, ... in which the "even" division is seldom possible. If a mother wants to divide "evenly" 9 apples among 3 children, there is no problem: each child gets 3 apples: $9 \div 3 = 3$. If, however, she wants to divide "evenly" 9 apples among 4 children, there is a problem (if we do not cut apples, i.e., stay within natural numbers). The solution is that each child gets 2 apples and 1 apple remains: $9 = 4 \times 2 + 1$. This situation is called in Mathematics division with remainder. The number to be divided is called dividend (9 in the last example), the number (of groups) we divide by is the divisor (4), the number of items in each group is the quotient (2), and what is left is the remainder (1). The Division Algorithm assures that such division with remainder is always possible. That means, given any two natural numbers: dividend and divisor, there exist unique natural numbers: quotient and remainder such that

dividend = divisor × quotient + remainder

and the remainder is less than the divisor. Now, we will explain how to divide with remainder using Maya numerals. We will follow the pattern explained in the chapter on Inca Mathematics; in fact, when describing division in Inca arithmetic we reversed the set-up used for multiplication, as there is no written record of Inca division method. The only difference is that in Inca arithmetic we use the decimal system, whereas in Maya—vigesimal. And this difference matters only when we trade.

First, let us reverse the multiplication example 4.2.5.

Example 4.2.6

Divide 5,456 by 62. Here 5456 =  and 62 = 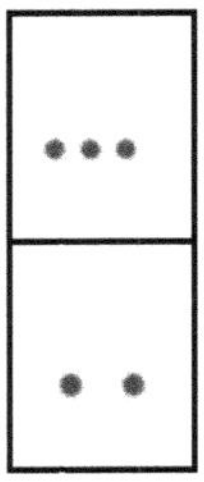

3 in 13 is 4 times, enter 4 in (1)

$4 \times 3 = 12$, enter 12 in (2)

$4 \times 2 = 8$, enter 8 in (3)

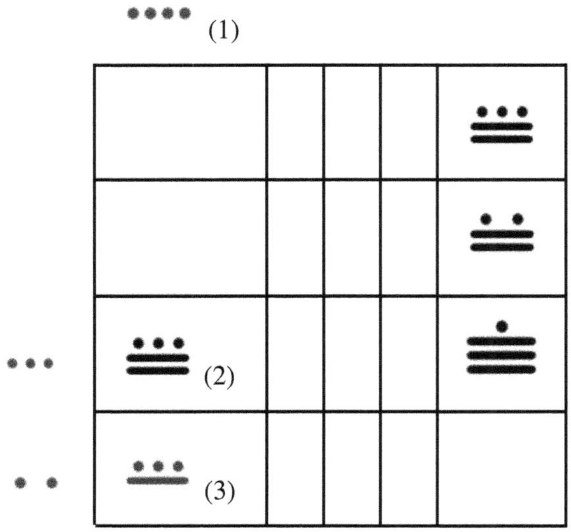

Now, find the differences $13 - 12 = 1$ and $12 - 8 = 4$ and enter them in (4) and (5), enter the last number in the dividend in (6).

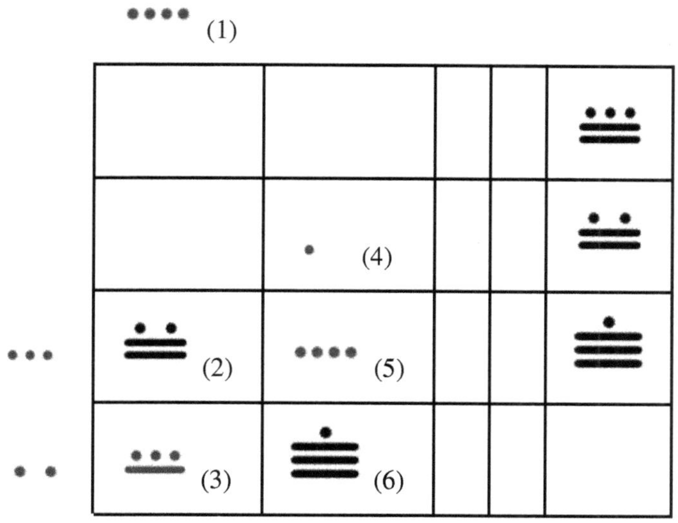

3 in 1 does not fit, 3 in 24 fits 8 times. Write 8 in (7)

$8 \times 3 = 24$, enter 24 in (8) and (9)

$8 \times 2 = 16$ enter 16 in (10).

Now find the differences $1 - 1 = 0$, $4 - 4 = 0$, and $16 - 16 = 0$ and enter them in (11), (12), and (13). This is the remainder.

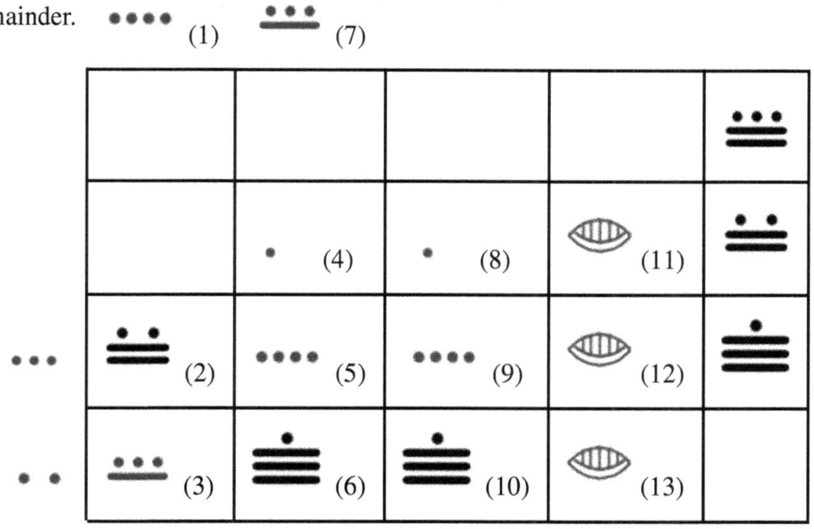

Let us work out another example.

Example 4.2.7

Suppose the dividend is 278, the divisor is 6. Translate into Maya numerals: $278 = 20 \times 13 + 18 =$ 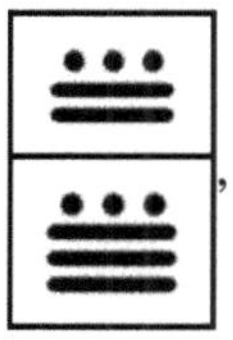,

$6 =$ 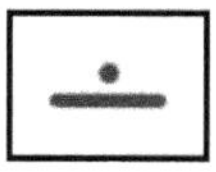

and set up the grid: 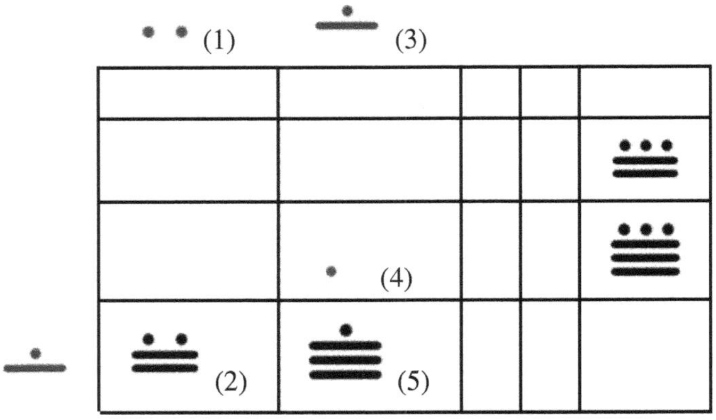

Fit 6 into 13 to get 2. Write 2 in (1), multiply 6 × 2 = 12, and enter 12 in (2). Subtract 13 − 12 = 1. Then bring the next digit 18 and fit 6 into 1 × 20 + 18 = 38 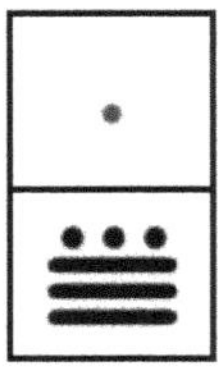 to get 6. Enter 6 in (3). Multiply the divisor 6 by 6 to get 36 and enter 36 in (4), (5). Subtract 38 - 36. The remainder is 2.

In conclusion, we get 278 = 6 × 46 + 2

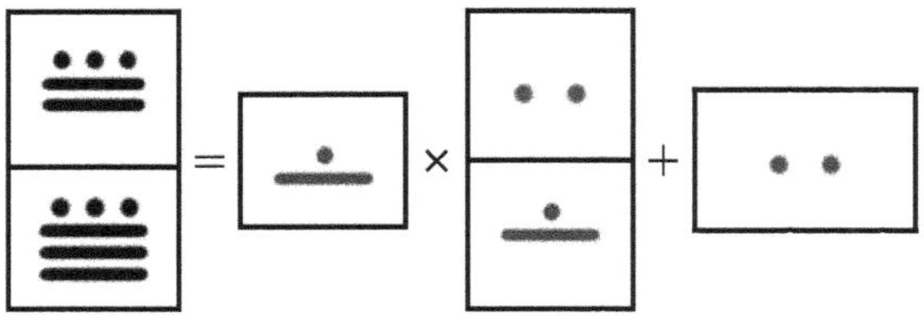

Example 4.2.8

Suppose the dividend is 2,643, divisor is 46. In Maya numerals: 2,643 = 6 × 20^2 + 12 × 20 + 3 = 

and 46 = 2 × 20 + 6 = 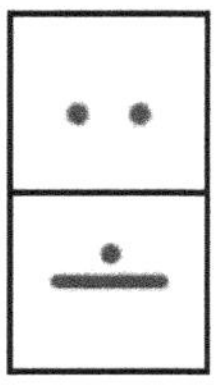

Let us set up the grid. 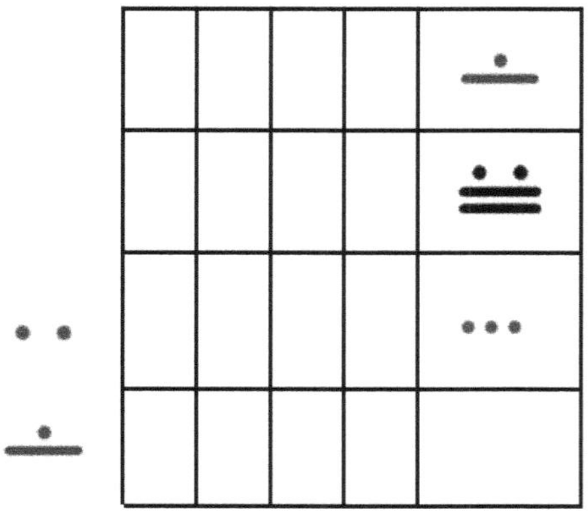

Like in case of Vedic division, we first take into account only the first digit of the divisor. Hence, we have to anticipate the development of calculation. The number 2 (the first digit of the divisor) goes into 6 (the first digit of the dividend) 3 times, but after subtraction $6-6$ we end up with 0 and it is impossible to take the next step. Therefore, we take 2 for the first digit of the quotient and enter it in (1).

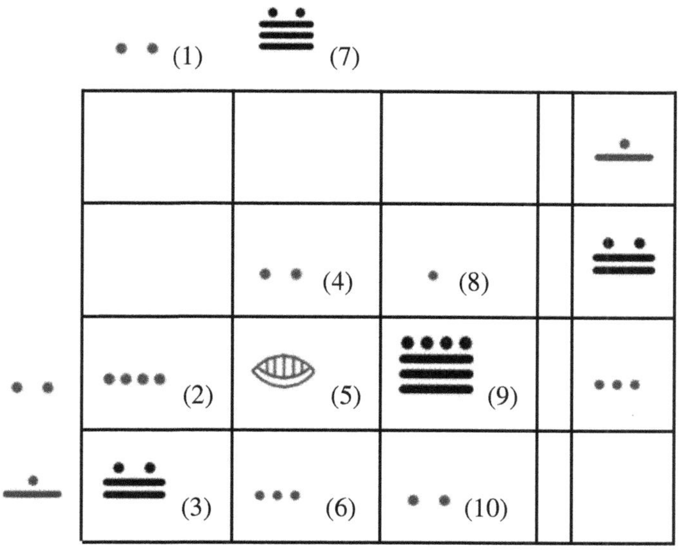

Multiply $2 \times 2 = 4$, enter 4 in (2)

$2 \times 6 = 12$, enter 12 in (3).

Now, find the differences: $6-4=2$, enter in (4), and $12-12=0$, enter in (5). Enter the last digit of the dividend in (6). Now we fit the divisor into the number in the second column 803.

Even though 2 goes into 2 once, 1 is not the next digit of the quotient. Just try 1 and you will see that in the subtraction in the next step you run into a negative number. Therefore, we fit the entire divisor 46 into 803. As a result we get 17, which is a Maya digit. Enter 17 in (7).

$17 \times 2 = 34 =$ 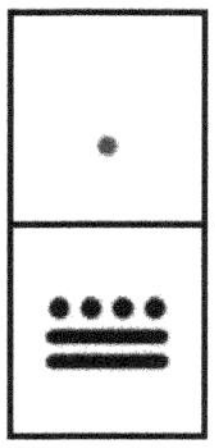 enter in (8), (9).

$17 \times 6 = 102 =$ 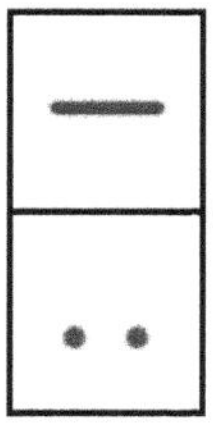 enter in (9), (10).

Now, we have to subtract 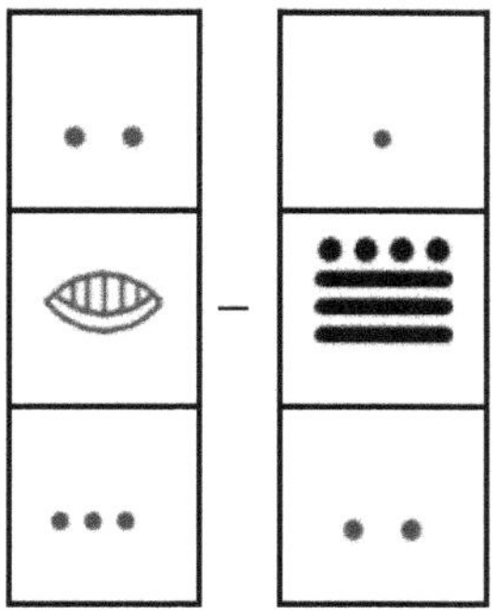

which requires trading:

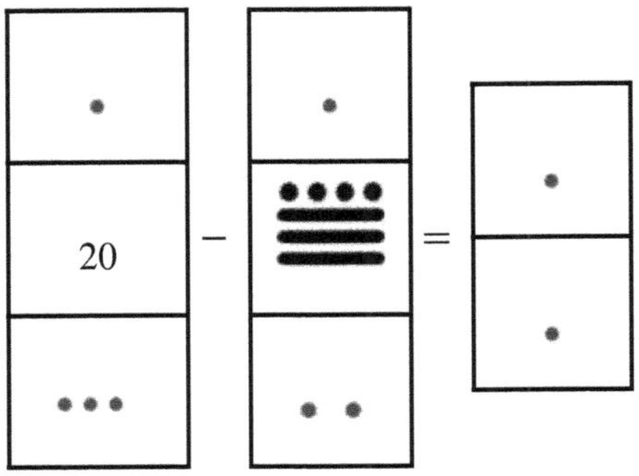

Hence, the remainder is 21. Therefore: $2{,}643 = 46 \times 57 + 21$

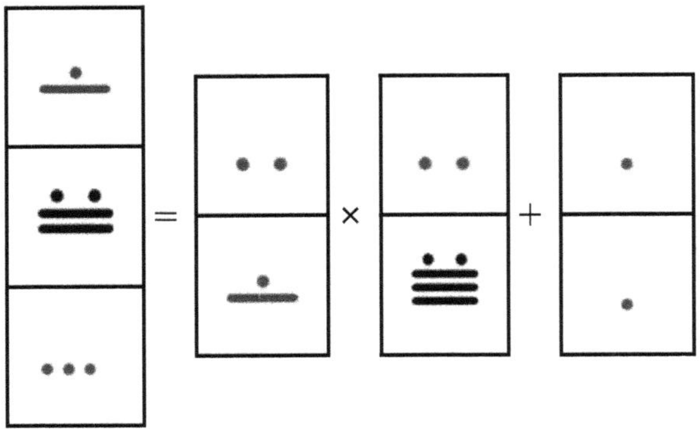

EXERCISES

1. Write the following numbers in Yucatec:

   a. 50 **b.** 70 **c.** 1,200
   d. 390 **e.** 2,000 **f.** 1,190

2. Write the following Yucatec Maya numbers in Hindu-Arabic:

   a. lahu ca bak **b.** ox pic **c.** can calab
   d. lahu ox calab **e.** ox tu y ho kal **f.** uaxac lahun tu y ho kal
   g. bolon lahun tu kal **h.** ho tu kal **i.** ho lahun tu y can kal
   j. ua lahun tu y ho kal

3. Convert into Maya numerals:

   a. 10 **b.** 27 **c.** 100
   d. 520 **e.** 1,000 **f.** 650
   g. 99 **h.** 12,521 **i.** 241,000

4. Convert the Hindu-Arabic numerals in exercise 3 into chronological/calendrical Maya numerals.
5. Convert into Hindu-Arabic numerals:

   a. 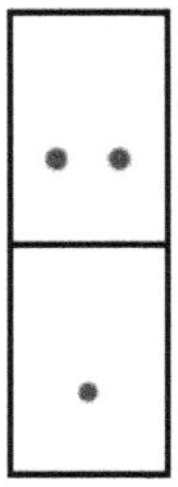 **b.** 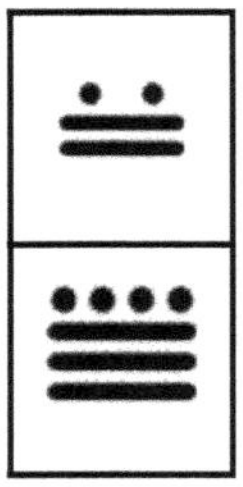 **c.** 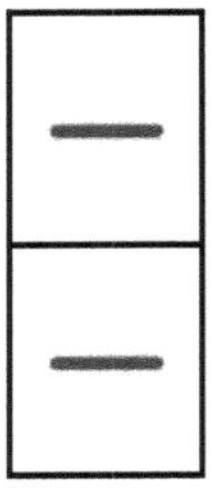 **d.** 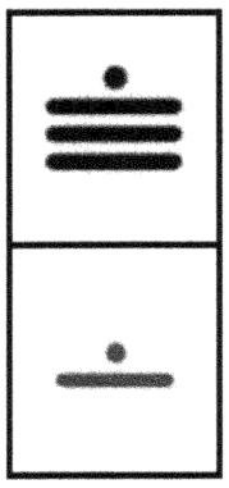 **e.** 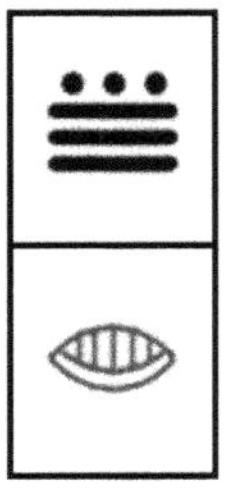

f. 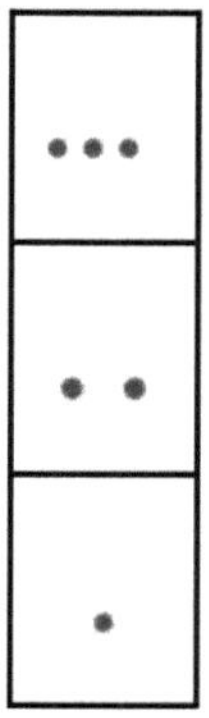

g. 

h. 

i. 

j. 

In problems 6 and 7 convert into Maya numerals, perform the operations in Maya numerals, and check in Hindu-Arabic numerals.

6. **a.** 40 + 55 **b.** 125 + 100 **c.** 400 + 261 **d.** 520 + 803
e. 458 + 883 **f.** 653 + 567 **g.** 1013 + 209 **h.** 602 + 227

7. **a.** 72 − 30 **b.** 514 − 433 **c.** 110 − 73 **d.** 211 − 95
e. 940 − 65 **f.** 1,008 − 525 **g.** 563 − 165 **h.** 1,306 − 612

8. The Maya Post-Classic period took place between 900 AD and 1519 AD. Use Maya symbols to write these dates and also the number of years the period lasted.

9. Maya astronomers calculated the Mars synodic period to be 780 days, write the number of days using Maya symbols.

10. In the Dresden Codex, the Maya wrote predictions of lunar and solar eclipses. Two of them were: July 11, 1991 and August 11, 1999. Write the years using Maya symbols.

11. The Maya Calendar Round consists of 18,980 days.

a. Each Tzolkin Calendar year has 260 days. Calculate the number of Tzolkin years in the Calendar Round using only Maya symbols to perform the division.

b. Each Haab Calendar year has 365 days. Calculate the number of Haab years in the Calendar Round using only Maya symbols to perform the division.

12. The Maya Metonic Cycle consists of 6,940 days which is equivalent to 235 lunar cycles. Using only Maya symbols calculate the number of days in a lunar cycle.

13. Figure 4.6 shows the middle section of page 44 from the Dresden Codex which contains a table of multiples of 78 using the chronological/calendrical system. In the lower half of the middle section, the list of Maya numbers are from left to right: 10 × 78 (here the bottom digit is zero in red), and in black: 5 × 78, 4 × 78, 3 × 78, 2 × 78, and 1 × 78. In the upper half of this middle section, are the 6, 7, 8, and 9 multiples of 78, together with the products 50 × 260 and 60 × 260.
Write all these products using Maya numbers using the chronological/calendrical system and compare your results with the Dresden Codex picture in Figure 4.6.

REFERENCES

Beltran de Santa Rosa, M.P. 1859. Arte de el idioma maya reducido a suscintas reglas y semilexicon yucateco. Merida, Mexico.

Calderon, Hector. 1966. La Ciencia Matematica de los Mayas, Mexico, D.F: Editorial Orion.

Coe, Michael D, and Mark Van Stone. 20013 *Reading the Maya Glyphs*, Thames & Hudson.

De Landa, Diego. 2005. *Relacion de las Cosas de Yucatan.* Mexico: Editorial San Fernando. (Version original 1566)

Heil, Nicholas. 2015. Unpublished Millersville University of Pennsylvania Honors Thesis.

Ifrah, Georges. 2000. *The Universal History of Numbers.* Wiley.

Morley, Sylvanus Griswold. 1975. *An Introduction to the Study of the Maya Hieroglyphs.* New York, NY: Dover Publications, Inc.

Stuart, David. 2011. *The Order of Days.* New York, NY: Harmony Books.

Thompson, John Eric Sydney. 1971. *Maya Hieroglyphic Writing*. Norman, OK: University of Oklahoma Press.

Tozzer, Alfred. *Landa's Relación de Cosas de Yucatan, Cambridge: Papers of the Peabody Museum of American Archaeology and Ethnology*. Vol. 18. Cambridge, MA: Harvard University.

Villacorta, Carlos, and Jose Antonio Villacorta. 1930. *The Dresden Codex, Drawings of the Pages and Commentary in Spanish*. Laguna Hills, CA: Aegean Park Press.

CHAPTER 5

The Calendars of the Maya

The Indians carried this letter wrapped up in their hair, and delivered it to Aguilar, and the ships, since the Indians took more time than they had expected, thought that they were dead and returned to the harbor of Cozumel. Cortés seeing that the Indians had not returned with the bearded men, set sail the next day. But that day one of the vessels sprung a leak and they were obliged to return to port, and as they were repairing her, Aguilar having received the letter, crossed in a canoe the strait between Yucatan and Cozumel, and as those of the fleet saw him they went to see who it was, and Aguilar asked them if they were Christians, and they replying that they were both Christians and Spaniards, he wept for joy, and kneeling down, he thanked God, and asked the Spaniards if the day was Wednesday.
(De Landa, 1566; Tozzer, 1941)

The Mesoamerican calendars were many and quite complex and there have been a good number of studies done to decipher them. By the arrival of Hernán Cortés in 1519 in what is today Mexico there were still 21 calendars in use while 4 of them were extincted: the Olmec, the Teotihuacán, the Yucuñudahui, and the Tolteca. (Edmonson, 1995)

Classical Maya civilization was highly advanced and developed many areas of science. Perhaps the most significant one was astronomy, which was important for agriculture and therefore, survival of their people. Based on astronomical observations, the Maya developed an elaborate system of calendars, among them the Tzolkin Calendar, the Haab Calendar, the Round Calendar which is a combination of the first two, and the Long Count. In this chapter, we study the calendars and discuss some relationships between the calendars from the mathematical standpoint.

It is quite worthwhile to mention that there was a correlation among the Maya and Aztec calendars; in general, there was great uniformity among the Mesoamerican calendars. (Caso, 1967) Some scholars believe that Stela A in Copan, Honduras, contains evidence that the Maya were able to calculate the year to an accuracy of 365.2422 days.

Four surviving Maya manuscripts (codices) have been found. These books have colored pages, inscribed on both sides and written on paper manufactured from the leaves of the maguey plant. The Maya name for the plant is ci, a plant that was also used to make the maguey liquor drunk by the Maya during their festivities. The codices' pages have been put together in folds and contain combinations of numbers, astrological and astronomical reckonings, information regarding wars, hurricanes, famines, and many other events.

The four known codices are:

1. The Dresden Codex at the Royal Library in Dresden.
2. The Paris Codex at the Biblioteque Nationale of Paris.
3. The Madrid Codex at Museo de América in Madrid.
4. The Grolier Codex in a Mexican vault, found in 1965, the authenticity of this codex is being disputed.

Many other codices existed as the missionary Diego de Landa described in his book "Relacion de las Cosas de Yucatan": *We found great number of books in these letters (the Maya writting system), but as they contained nothing that did not savor of superstition and lies of the devil we burnt them all, at which the natives grieved most keenly and were greatly pained.*

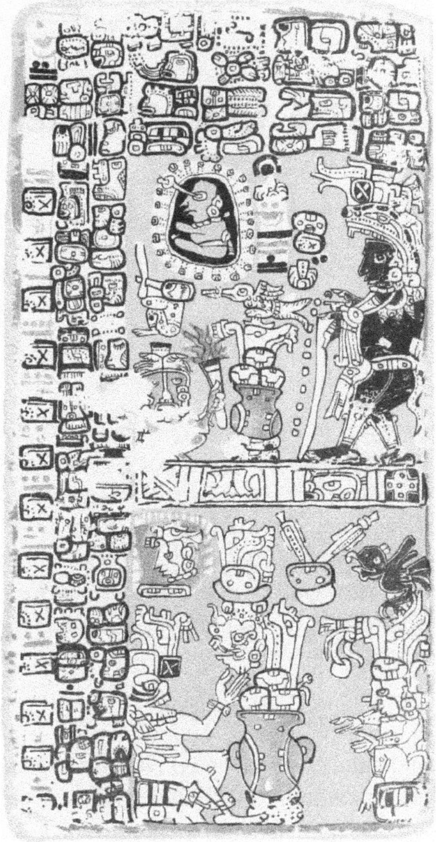

From *A Study of the Manuscript Troano* by Cyrus Thomas, 1882, Washington Government Printing Office.

Figure 5.1: Madrid Codex, Plate XXIII, a Maya astronomer

Figure 5.1 can be found on Plate XXIII of the Madrid Codex, the longest of the surviving Maya manuscripts. The upper part of the drawing depicts a Maya astronomer observing the stars in the darkness of the night. Notice his elongated eye reaching out to the celestial sphere (Gendrop, 1970).

El Caracol, Spanish for "*the snail*" is believed to have been a Maya observatory located in Chichen Itza "*the wells of the Itza*", Yucatán, Mexico 850 – 1000AD. The only three remaining windows of the two-story circular building offered a view to watch astronomical events, especially the planet Venus, which is also thoroughly studied in the codices. Venus, above all planets visible to the naked eye, was of upmost importance for the Maya astronomers, it was called *chac ek,* red star in Yucatec among other names (Figure 5.2).

5.1 The Calendars

The earliest Maya Long Count date yet discovered was carved on Stela 2 at Chiapa de Corzo (in modern day Chiapas). It records the Long Count date 7. 16. 3. 2. 13, in the year 36 BC. (Maya Exploration Center). A detailed description of Long Count dates is presented later in this chapter.

In Tikal, the oldest, and contemporaneous with its carving, Maya known date was found on a stela, known as Stela 29, and it corresponds to the year 292 of our era.

The monument shows a ruler holding a double-headed serpent bar with supernaturals emerging from the open jaws, an ancestor floats overhead (FAMSI). Its Long Count date is 8. 12. 14. 8. 15 corresponding to the Round Calendar date 13 Men 3 Zip (Figure 5.3). A Round Calendar date is a combination of the Tzolkin and Haab Maya calendars that are outlined below.

Courtesy of Ximena Catepillán, 2006

Figure 5.2: El caracol

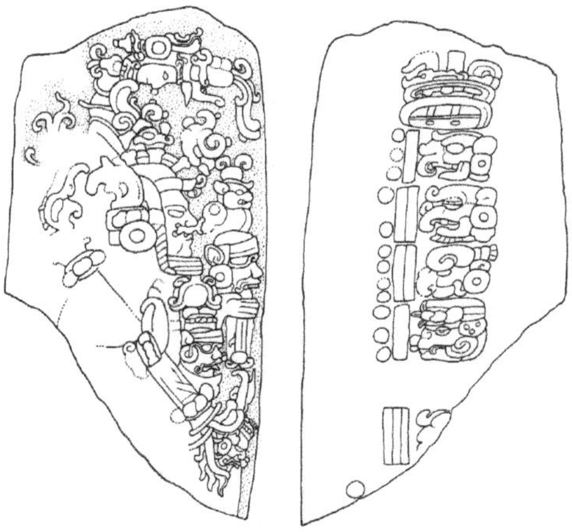

Drawing by Linda Schele © David Schele

Figure 5.3: Stela 29 (Schele, 2000)

Tzolkin Calendar

The 260-day Tzolkin, or sacred calendar, is the oldest calendar cycle known in Mesoamerica, dating back to at least 600 BC. While some scholars are still searching for an astronomical basis for this cycle, some agree it was based on the nine month human gestation period. The Maya name for the calendar is not known for certain; but in Mayanist literature is commonly called the Tzolkin ("the sequence of days," "the naming-in-order of the days") (Lounsbury, 1978).

The calendar is still in use by some of the shamans in the Highlands of Guatemala for divination and ceremonial purposes. The picture in Figure 5.4 was taken in 2007 in Chichicastenango, Guatemala, during a shamans' ceremony.

The Tzolkin Calendar, also called the Tonalamatl or Sacred Calendar, uses 20 names of gods (veintena) with 13 days (trecena) attached. The "veintena" and "trecena" names are convenient Spanish terms used in the literature. The names of the gods in Yucatec (not exact transcription) can be seen in Table 5.1. Figure 5.5 illustrates the Tzolkin 260-day count, in which 20 gods intermesh with 13 numbers.

An entire Tzolkin calendar cycle is delineated in Table 5.2. To follow the sequence of days begin with 1 Imix, using the number 1 in the leftmost column of numbers and the first god, 2 Ik, 3 Akbal, 4 Kan, ..., 7 Ahau, now continue reading down the second column of numbers, 8 Imix, 9 Ik, 10 Akbal, ..., and so on until you reach the last day of the calendar 13 Ahau.

Courtesy of Ximena Catepillán, 2007

Figure 5.4: Shamans' ceremony

Table 5.1: The Tzolkin gods

Tzolkin gods
Imix
Ik
Akbal
Kan
Chicchan
Cimi
Manik
Lamat
Muluc
Oc
Chuen
Eb
Ben
Ix
Men
Cib
Caban
Etznab
Cauac
Ahau

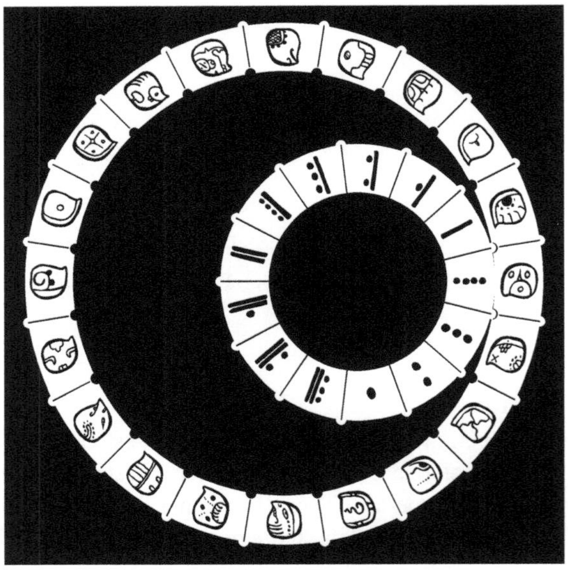

Figure 5.5: Tzolkin calendar wheels

Table 5.2: The Tzolkin Calendar

Tzolkin Calendar													
Imix	1	8	2	9	3	10	4	11	5	12	6	13	7
Ik	2	9	3	10	4	11	5	12	6	13	7	1	8
Akbal	3	10	4	11	5	12	6	13	7	1	8	2	9
Kan	4	11	5	12	6	13	7	1	8	2	9	3	10
Chicchan	5	12	6	13	7	1	8	2	9	3	10	4	11
Cimi	6	13	7	1	8	2	9	3	10	4	11	5	12
Manik	7	1	8	2	9	3	10	4	11	5	12	6	13
Lamat	8	2	9	3	10	4	11	5	12	6	13	7	1
Muluc	9	3	10	4	11	5	12	6	13	7	1	8	2
Oc	10	4	11	5	12	6	13	7	1	8	2	9	3
Chuen	11	5	12	6	13	7	1	8	2	9	3	10	4
Eb	12	6	13	7	1	8	2	9	3	10	4	11	5
Ben	13	7	1	8	2	9	3	10	4	11	5	12	6
Ix	1	8	2	9	3	10	4	11	5	12	6	13	7
Men	2	9	3	10	4	11	5	12	6	13	7	1	8
Cib	3	10	4	11	5	12	6	13	7	1	8	2	9
Caban	4	11	5	12	6	13	7	1	8	2	9	3	10
Etznab	5	12	6	13	7	1	8	2	9	3	10	4	11
Cauac	6	13	7	1	8	2	9	3	10	4	11	5	12
Ahau	7	1	8	2	9	3	10	4	11	5	12	6	13

As a result 260 dates are formed, since the least common multiple of 20 and 13 is 260, which constitute 1 Tzolkin year. The year begins with 1 Imix and ends with 13 Ahau. Figure 5.6 shows the 20 Tzolkin gods symbols for the codices.

Haab Calendar

The 365-day Haab calendar is the one most similar to the Christian calendar. With 365 days in its count, it is obviously based on solar observations. Haab is a Yucatec name and "the apparent derivation of the term implies a reference to the seasonal year (the "rains" being for the count of years in the tropics what "winters" have been in the temperate zones)" (Lounsbury, 1978).

The Haab Calendar, is also called Vague, Solar, or Civil Calendar. The name Vague is sometimes used "since although the ancient calendar specialists were fully aware that the actual length of the year was about 365 $\frac{1}{4}$ days, they never intercalated leap days as we do, or anything similar." (Coe and Van Stone, 2001). This calendar uses 18 month names (gods) of 20 days each to make a year of 360 days plus five days or Uayeb. These five days were called "the nameless days" and were considered unlucky. The Maya made public offerings to their gods during these five days, they did not

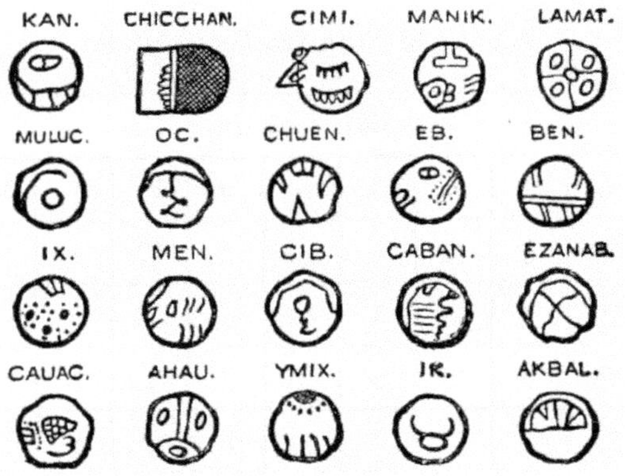

Courtesy of Library of Congress

Figure 5.6: The Tzolkin gods

cleanse themselves. Moreover, they refrained from doing hard work for fear that some misfortune might had fallen upon them. This calendar was used also by the Zapotecos, Mixtecos, Totonacos, Tarascos, Matlatzincas, Otomies and the Nahuas. It is possible that they knew of the lenght of the year to be a fraction more than 365, the correction was not done in the calendar but they had the so called "anniversary dates" to fix the problem (Caso, 1967; Teeple, 1931).

The Yucatec names of the gods (not exact transcription) are: Pop, Uo, Zip, Zotz, Tzec, Xul, Yaxkin, Mol, Chen, Yax, Zac, Ceh, Mac, Kankin, Muan, Pax, Kayab, Cumku, and Uayeb (Table 5.3 and Figure 5.7).

In order to prepare for the celebration of the first day of Pop, the first day of the New Year, the priests and the nobility would fast for at least 13 days prior to this important event. The Maya assembled to feast with all the offerings of food and wine newly fermented, in honor of their gods.

Figure 5.8 shows the Haab gods symbols used on the stone inscriptions.

To each god the numbers 0, 1, . . . , 19 are attached as in our Gregorian calendar, but for the 0th day of each month they did not use the symbol 0 but a glyph called "seating", *chum* in Yucatec. This *chum*—god date was really the last day of the previous month. For example in the sequence of dates ..., 18 Cumku, 19 Cumku, 0 Uayeb, 1 Uayeb, 2 Uayeb, 3 Uayeb, 4 Uayeb, 0 Pop, 1 Pop, 2 Pop... the date 0 Uayeb is the last day of the month Cumku, and the date 0 Pop is the last day of the mini-month Uayeb, these dates are the "seatings" of the gods for the following months. In today' s glyph representations we see a symbol for zero in lieu of the *chum* glyph, but the Maya did not use zero for the last days of each month. See the *chum*—Pop date in Figure 5.9.

The Uayeb month of five days was considered for some an unlucky month, planning events during these days was strongly discouraged. "The name of the final month of the year, the five-day period, Uayeb, means "bed," a sleeping place. It is the "seating" of the year. It is also noted by some scholars that this five-day mini-month was spent conducting special ceremonies, as recorded in the Madrid Codex with a description of the New year's rituals.

As a result of adding the Uayeb mini-month to the 360 days, 365 dates are formed, which constitute 1 Haab year. The 365-day Haab calendar is delineated in Table 5.4.

Round Calendar

The Maya merged the Tzolkin and Haab calendars into a superior cycle known as the Calendar Round, a calendar that was common in all Mesoamerica (Caso, 1967). The combined calendar consists of 18,980 days since $260 = 13 \times 2^2 \times 5$ and $365 = 73 \times 5$ and therefore the least common multiple of 260 and 365, $\text{lcm}(260, 365) = 13 \times 2^2 \times 73 \times 5 = 18,980$ which corresponds to 73 Tzolkin years or 52 Haab years, $73 \times 260 = 52 \times 365 = 18,980$.

The Round Calendar date gives us the exact day within one cycle of 18,980 days, but we cannot locate the date in time as the Round calendar cycles are not numbered (like our years, e.g., 2014). In a Calendar Round date the Tzolkin date is written first and then the Haab date, thus there are four entries.

Table 5.3: The Haab gods

Haab gods
Pop
Uo
Zip
Zotz
Tzec
Xul
Yaxkin
Mol
Chen
Yax
Zac
Ceh
Mac
Kankin
Muan
Pax
Kayab
Cumku
Uayeb

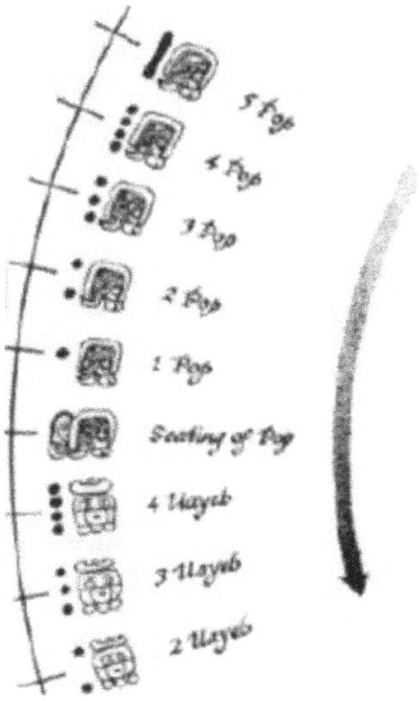

Figure 5.7: Haab calendar wheel

Figure 5.8: Haab gods

Illustration by Mark Van STone from *Reading the Maya Glyphs* by Michael D. Coe, Thames & Hudson, London and New York.

Figure 5.9: Chum - Pop or 0 Pop (Coe and Van Stone, 2001)

Table 5.4: The Haab Calendar

Haab Calendar																				
Pop	0	1	2	3	4	5	6	7	8	9	10	11	12	13	14	15	16	17	18	19
Uo	0	1	2	3	4	5	6	7	8	9	10	11	12	13	14	15	16	17	18	19
Zip	0	1	2	3	4	5	6	7	8	9	10	11	12	13	14	15	16	17	18	19
Zotz	0	1	2	3	4	5	6	7	8	9	10	11	12	13	14	15	16	17	18	19
Tzec	0	1	2	3	4	5	6	7	8	9	10	11	12	13	14	15	16	17	18	19
Xul	0	1	2	3	4	5	6	7	8	9	10	11	12	13	14	15	16	17	18	19
Yaxkin	0	1	2	3	4	5	6	7	8	9	10	11	12	13	14	15	16	17	18	19
Mol	0	1	2	3	4	5	6	7	8	9	10	11	12	13	14	15	16	17	18	19
Chen	0	1	2	3	4	5	6	7	8	9	10	11	12	13	14	15	16	17	18	19
Yax	0	1	2	3	4	5	6	7	8	9	10	11	12	13	14	15	16	17	18	19
Zac	0	1	2	3	4	5	6	7	8	9	10	11	12	13	14	15	16	17	18	19
Ceh	0	1	2	3	4	5	6	7	8	9	10	11	12	13	14	15	16	17	18	19
Mac	0	1	2	3	4	5	6	7	8	9	10	11	12	13	14	15	16	17	18	19
Kankin	0	1	2	3	4	5	6	7	8	9	10	11	12	13	14	15	16	17	18	19
Muan	0	1	2	3	4	5	6	7	8	9	10	11	12	13	14	15	16	17	18	19
Pax	0	1	2	3	4	5	6	7	8	9	10	11	12	13	14	15	16	17	18	19
Kayab	0	1	2	3	4	5	6	7	8	9	10	11	12	13	14	15	16	17	18	19
Cumku	0	1	2	3	4	5	6	7	8	9	10	11	12	13	14	15	16	17	18	19
Uayeb	0	1	2	3	4															

An example of a Round Calendar date is 4 Ahau 8 Cumku, in which 4 Ahau is the Tzolkin date and 8 Cumku is the Haab date. It takes 18,980 days until this date shows up again. Notice a sample sequence of dates below:

Tzolkin Calendar		Haab Calendar		Gregorian Calendar (an example)
4	Ahau	8	Cumku	March 24, 1980
5	Imix	9	Cumku	March 25, 1980
6	Ik	10	Cumku	March 26, 1980
7	Akbal	11	Cumku	March 27, 1980
8	Kan	12	Cumku	March 28, 1980
9	Chicchan	13	Cumku	March 29, 1980
10	Cimi	14	Cumku	March 30, 1980
11	Manik	15	Cumku	March 31, 1980
12	Lamat	16	Cumku	April 1, 1980
13	Muluc	17	Cumku	April 2, 1980
1	Oc	18	Cumku	April 3, 1980
2	Chuen	19	Cumku	April 4, 1980
3	Eb	0	Uayeb	April 5, 1980
4	Ben	1	Uayeb	April 6, 1980
5	Ix	2	Uayeb	April 7, 1980
6	Men	3	Uayeb	April 8, 1980
7	Cib	4	Uayeb	April 9, 1980
8	Caban	0	Pop	April 10, 1980
9	Etznab	1	Pop	April 11, 1980
	...			

Figure 5.10 depicts the Round calendar in action, on the left the Tzolkin Calendar in which the outer wheel is representing the 20 gods and the inner wheel the 13 day numbers. On the right, we see part of the big Haab calendar wheel.

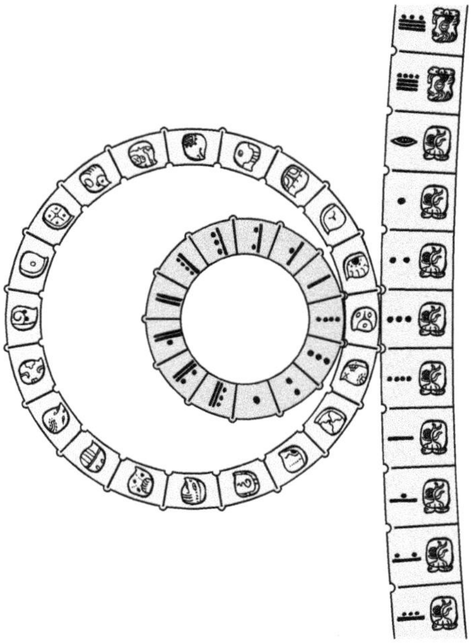

Figure 5.10: The Round Calendar

Examples

Given the Round Calendar date 6 Ahau 13 Yaxkin, August 13, 2014.

a. What was the Round Calendar date 3 days later?
To find the date we need to add 3 to 6 (the day number in the Tzolkin calendar), also we move forward 3 gods from Ahau, and we add 3 to 13 (the day number in the Haab calendar).

6	Ahau	13	Yaxkin
+ 3	move 3 gods forward	+ 3	
9	Akbal	16	Yaxkin

Therefore the Round Calendar date is 9 Akbal 16 Yaxkin.

b. What was the date 9 days later?
To find the date we need to add 9 to 6, the day number, obtaing 15 and since we have only 13-day numbers, the number 15 corresponds to day number 2 (think of the numbers 1 through 13 on a clock face). We also move forward 9 gods from Ahau (see Table 5.1) and obtain god Muluc. Thus the Tzolkin date is 2 Muluc. To find the Haab date we add 9 to 13, the Haab day number, obtaining 22 which corresponds to day number 2 (think of the numbers 0 through 19 on a clock face). However, the Haab god is not Yaxkin because we reached 19 Yaxkin, the last Yaxkin date, and then we have 0 Mol (day 20)(the seating of the Yaxkin month) followed by 1 Mol (day 21) and then 2 Mol (day 22).

6	Ahau	13	Yaxkin	
+ 9	move 9 gods forward	+ 9		
2	Muluc	2	Mol	the god after Yaxkin

The Round Calendar date is 2 Muluc 2 Mol.

c. What was the date 15 days earlier? For the Tzolkin number calculation imagine the numbers 1 through 13 on a clock face, and move backwards 15 times starting at 6. To find the Tzolkin god move also backwards 15 places beginning with the god Ahau. Similarly for the Haab number day, move backwards 15 places on a 0 through 19 clock face, here we obtain 18 and we reach the god Xul, the god before Yaxkin.

To understand this process imaging that today is July 13 and you would like to know the date 15 days earlier. First, you would move 15 places backwards beginning with 13, the day number would be 28 and the month June, the month before July.

6	Ahau	13	Yaxkin	
– 15	move backwards 15 gods	– 15		
4	Chicchan	18	Xul	the god before Yaxkin

Hence, the Round Calendar date is 5 Chicchan 18 Xul.

What about leap years?

The Mesoamerican calendars did not have leap years. Hence, the Haab calendar gradually was falling apart from the tropical year. The astronomers were aware of the problem and even though they took care of it in their astronomical observations, by calculating the accumulated error, they did not modify the calendar. The astronomers of Kaminaljuyú of the Highlands of Guatemala performed the right computation by decreasing the calendar by one year every 29 calendar rounds of 52 years (18,980 days) because $29 \times 18,980 = 550,420$ days = 1,508 Haab years (365-day) = 1,507 tropical years (365.2422-day). The Kaminaljuyú priests knew this around 433 BC (Edmonson, 1995).

Correlation is a number of days we add to the Maya Long Count dates to obtain the Julian Calendar date used in astronomy, the Julian calendar (365.25 day calendar) was in use by the Europeans prior to the adoption of the more accurate Gregorian calendar (365.2425 day calendar) in use today. There is a surprisingly large uncertainty in the value of the correlation, yielding a shift between both calendars (and thus between the history of Maya and of our world) of typically several hundred years (Bohm, 2013).

Long Count Dates

The Long Count, for which we do not know the Maya name, is commonly considered the Maya's linear count of days. In truth, it is yet another cycle, but its great length of 5,126 years makes it essentially a linear count through all of Maya history (Maya Exploration Center).

The Maya used the Long Count dates to mark events, which ocurred over long periods of time. The shortest period was a *kin,* a day for us, and 20 *kins* was a *uinal*, and then a *tun* consisted of 18 *uinals* or 360 *kins, a katun* had 20 *tuns* or 7,200 *kins,* and lastly, a *baktun* consisted of 20 *katuns* or 144,000 *kins* (Figure 5.11).

In summary:

1 uinal	=	20 kins		
1 tun	=	18 uinals	=	360 kins
1 katun	=	20 tuns	=	7,200 kins
1 baktun	=	20 katuns	=	144,000 kins

A Long Count date typically begins wilh a special combination of glyphs named by the epigraphers the Initial Series Introductory Glyph (ISIG). The ISIG glyph contains four fixed parts, one of them being the "tun" glyph, and a variable part usually in the central part of the ISIG. There are different possible meanings of this variable part of the glyp according to the literature. One of them states that it is in relation to the Haab month, and hence the name "Patron of the Month" (Houston et al., 2001). The Maya name for this glyph is still unknown. Figure 5.12 illustrates a sample of six drawings used to represent the ISIG.

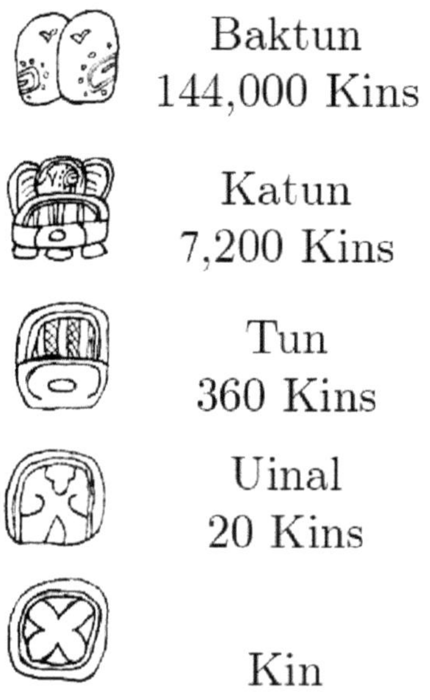

Drawing by Ximena Catepillán, 2015

Figure 5.11: Periods of the Long Count

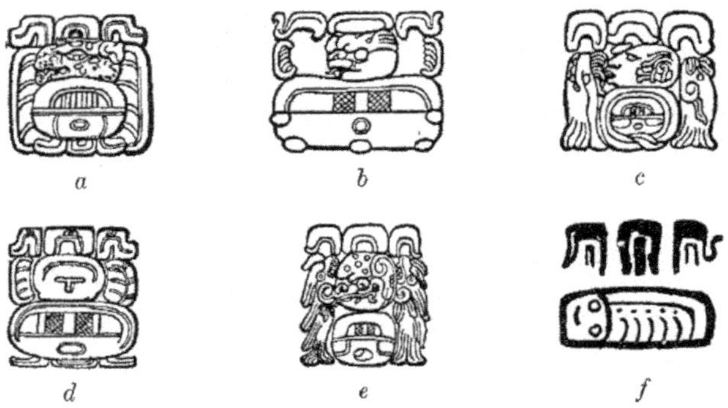

Figure 5.12: Initial Series Introductory Glyphs (Morley, 1947a, b)

For example, Figure 5.13 depicts the Long Count date 12. 19. 15. 15. 8.

First we have the Initial Series Introductory Glyph ISIG, the symbol identifies the date as a Long Count date. Then we read the rows from left to right as we move down. Therefore, we have:

12 Baktuns: The Maya number 12 with the symbol of baktun.

19 Katuns: The Maya number 19 with the symbol for katun.

15 Tuns: The Maya number 15 with the symbol for tun.

15 Uinals: The Maya number 15 with the symbol for uinal.

8 Kins: The Maya number 8 with the symbol for kin.

7 Lamat: The Maya number 7 with the Tzolkin god glyph Lamat is the Tzolkin calendar date.

11 Ceh: The Maya number 11 with the Haab god glyph Ceh is the Haab calendar date.

The last glyph, in the bottom right of Figure 5.13, represents one of the nine Lords of the Night of the Maya underworld known as Xibalba, the "Place of Fright". These nine gods ruled over the nights in sequence and their depictions are ubiquitous in the Maya manuscripts. However, none of their representations have the names of these deities, as opposed to the Aztec group of nine lords, but there are enough references in the manuscripts that portray them as a group of nine gods with the Maya name *Bolon-ti-Ku,* "the nine gods". Epigraphers have named them gods G1 through G9. The glyph in the figure below portrays the Lord of the Night G2, the steps to find the deity number are given later in this chapter (Figure 5.13).

The Gregorian date for this Long Count date is November 20, 2008, you can use the Maya Exploration Center Calendar Converter site to check this and other dates.

Another example of a Long Count date can be found on a polished thin plaque of green jade, known as the Leyden Plaque, after the Holland' s city in whose museum is preserved. The pre-classic period plaque, found in Tikal, Guatemala, has the shape of a miniature stela with a human figure on one side and on the other side a Long Count date that reads 8. 14. 3. 1. 12. The date is preceded by an initial series introductory glyph and also depicts the Tzolkin and Haab dates (Figure 5.14). The date corresponds to the year 320 AD. The long count date tells us that there have been

$8 \times 144,000 + 14 \times 7,200 + 3 \times 360 + 1 \times 20 + 12 = 1,253,912$ kins since the beginning of the last Maya era. An era or creation of 13 baktuns has $13 \times 144,000 = 1,872,000$ kins, which is approximately 5,128 Haab years.

Before we perform computations among Long Count dates we will ilustrate how to convert a number of days (kins) into Long Count date, this will prove later to be a very useful tool to have.

How to convert a number of kins into Long Count date:

The oldest, and contemporaneous with its carving, Maya known date was found in Tikal, Guatemala, on a stela known as Stela 29, and it corresponds to the year 292 AD. The number of kins (days) corresponding to the Long Count date is 1,243,615 kins (Figure 5.15).

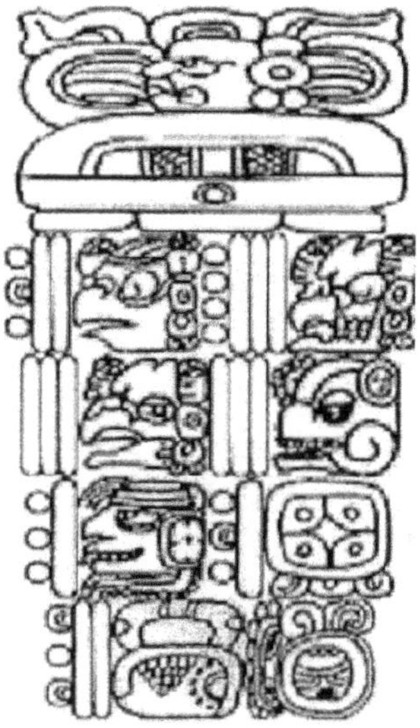

Figure 5.13: Long Count date 12. 19. 15. 15. 8 (Hartley, 2002)

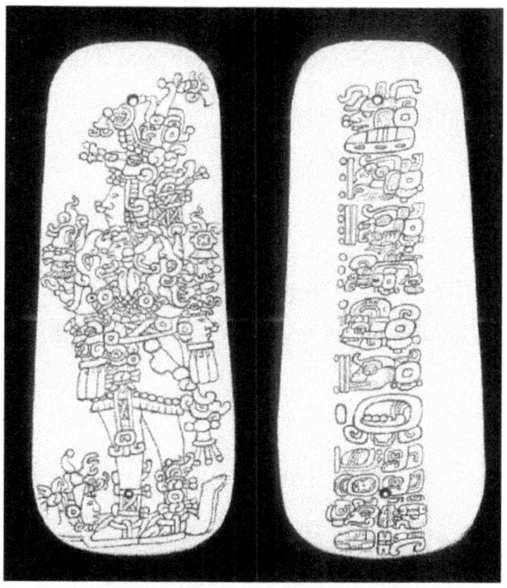

Figure 5.14: Leyden Plaque, drawing by Coe (1967)

To find the Long Count date associated to this number of kins, we use the division algorithm and following steps:

Step 1:
Find the number of baktuns by dividing the number of kins by 144,000 the number of days in 1 baktun.

$1,243,615 \div 144,000 = 8,636$, therefore the number of baktuns is 8.

The remainder in this division is 91,615 since $8 \times 144,000 = 1,152,000$ and $1,243,615 - 1,152,000 = 91,615$.

Step 2:
Find the number of katuns by dividing the remainder 91,615 by 7,200 the number of days in 1 katun.

$91,615 \div 7,200 = 12.724$, therefore the number of katuns is 12.

The remainder in this division is 5,215 since $12 \times 7,200 = 86,400$ and $91,615 - 86,400 = 5,215$.

From *Expedition Magazine, Vol. 2, Issue 2, Winter 1960* by Edwin M. Shook, Author and William R. Coe, Illustrator/Photographer.

Figure 5.15: Stela 29 (Coe 1967)

Step 3:
Find the number of tuns by dividing the remainder by 360, the number of days in 1 tun.

$5,215 \div 360 = 14.486$, therefore the number of tuns is 14.

The remainder in this division is 175 since $14 \times 360 = 5,040$ and $5,215 - 5,040 = 175$.

Step 4:
Finally to find the number of uinals we divide the remainder 175 by 20, the number of kins in 1 uinal.

$175 \div 20 = 8.75$, therefore the number of uinals is 8, and the remainder in this last division, which is 15, will be the last entry in the Long Count date.

$8 \times 20 = 160$ and $175 - 160 = 15$

Thus the Long Count date is 8. 12. 14. 8. 15, verify this date against the date in Figure 5.3.

Operations among Long Count dates
The Long Count date 0. 0. 0. 0. 0 corresponds to the Round Calendar date 4 Ahau 8 Cumku of the first cycle of 18,980 days. We will find the Long Count date of the beginning of the 98th cycle. In this case, the number of days elapsed since the beginning of the first cycle will be $18,980 \times 98 = 1,860,040$ days. Thus, if we convert this number of days into Long Count date we will have the desired date.

To find the Long Count date associated to this number of kins, we use the steps shown for Stela 29:

Step 1:
Find the number of baktuns by dividing the number of kins by 144,000 the number of days in 1 baktun.

$1,860,040 \div 144,000 = 12.917$, therefore the number of baktuns is 12.

The remainder in this division is 132,040 since $12 \times 144,000 = 1,728,000$ and $1,860,040 - 1,728,000 = 132,040$.

Step 2:
Find the number of katuns by dividing the remainder 132,040 by 7,200 the number of days in 1 katun.

$132,040 \div 7,200 = 18.339$, therefore the number of katuns is 18.

The remainder in this division is 2,440 since $18 \times 7,200 = 129,600$ and $132,040 - 129,600 = 2,440$.

Step 3:
Find the number of tuns by dividing the remainder by 360, the number of days in 1 tun.

$2,440 \div 360 = 6.778$, therefore the number of tuns is 6.

The remainder in this division is 280 since $6 \times 360 = 2,160$ and $2,440 - 2,160 = 280$.

Step 4:
Finally to find the number of uinals we divide the remainder 280 by 20, the number of kins in 1 uinal.

$280 \div 20 = 14$, therefore the number of uinals is 14, and the remainder in this last division, which is 0, will be the last entry in the Long Count date. Thus the Long Count date is 12. 18. 6. 14. 0. The number of kins corresponding to 12. 18. 6. 14. 0 can also be calculated as $12 \times 144,000 + 18 \times 7,200 + 6 \times 360 + 14 \times 20 + 0 = 1,860,040$.

Notice in the sequence of dates below how the corresponding Long Count dates are changing.

Round Calendar date				First Cycle					98th cycle of 18,980 days				
4	Ahau	8	Cumku	0.	0.	0.	0.	0	12.	18.	6.	14.	0
5	Imix	9	Cumku	0.	0.	0.	0.	1	12.	18.	6.	14.	1
6	Ik	10	Cumku	0.	0.	0.	0.	2	12.	18.	6.	14.	2
7	Akbal	11	Cumku	0.	0.	0.	0.	3	12.	18.	6.	14.	3
8	Kan	12	Cumku	0.	0.	0.	0.	4	12.	18.	6.	14.	4
9	Chicchan	13	Cumku	0.	0.	0.	0.	5	12.	18.	6.	14.	5
10	Cimi	14	Cumku	0.	0.	0.	0.	6	12.	18.	6.	14.	6
11	Manik	15	Cumku	0.	0.	0.	0.	7	12.	18.	6.	14.	7
12	Lamat	16	Cumku	0.	0.	0.	0.	8	12.	18.	6.	14.	8
13	Muluc	17	Cumku	0.	0.	0.	0.	9	12.	18.	6.	14.	9
1	Oc	18	Cumku	0.	0.	0.	0.	10	12.	18.	6.	14.	10
2	Chuen	19	Cumku	0.	0.	0.	0.	11	12.	18.	6.	14.	11
3	Eb	0	Uayeb	0.	0.	0.	0.	12	12.	18.	6.	14.	12
4	Ben	1	Uayeb	0.	0.	0.	0.	13	12.	18.	6.	14.	13
5	Ix	2	Uayeb	0.	0.	0.	0.	14	12.	18.	6.	14.	14
6	Men	3	Uayeb	0.	0.	0.	0.	15	12.	18.	6.	14.	15
7	Cib	4	Uayeb	0.	0.	0.	0.	16	12.	18.	6.	14.	16
8	Caban	0	Pop	0.	0.	0.	0.	17	12.	18.	6.	14.	17
9	Etznab	1	Pop	0.	0.	0.	0.	18	12.	18.	6.	14.	18
10	Cauac	2	Pop	0.	0.	0.	0.	19	12.	18.	6.	14.	19
11	Ahau	3	Pop	0.	0.	0.	1.	0	12.	18.	6.	15.	0
12	Imix	4	Pop	0.	0.	0.	1.	1	12.	18.	6.	15.	1
	...												

The Long Count date 12. 18. 6. 14. 0 also corresponds to the Round Calendar date 4 Ahau 8 Cumku, March 24, 1980 in our Gregorian calendar. The next Round Calendar date 4 Ahau 8 Cumku will occur in the year 2032 and the Long Count date will be 13. 0. 19. 9. 0 since the Round Calendar cycle consists of 18,980 days or 0. 2. 12. 13. 0 in Long Count form (it is a Distance number) and if we add the two dates we can confirm the answer.

	12.	18.	6.	14.	0	1,860,040 days	4 Ahau	8 Cumku	March 24, 1980
+	0.	2.	12.	13.	0	18,980	Distance number		
	13.	0.	19.	9.	0	1,879,020	4 Ahau	8 Cumku	March 11, 2032

Adding and Subtracting Long Count dates:
Example 5.1.1
If we add 1 kin (0. 0. 0. 0. 1) to the date 12. 19. 19. 17. 19 presented below, we obtain 13. 0. 0. 0. 0. This is because in the last entry (the kins entry) we obtain 20 kins, which corresponds to 1 uinal, therefore we have now 18 uinals in the uinals entry. Moreover, since 18 uinals = 1 tun we add this tun to the 19 tuns we had in the tuns entry obtaining 20 tuns, but 20 tuns = 1 katun, thus we add this katun to the 19 katuns we had in the katuns entry obtaining 20 katuns, and since 20 katuns = 1 baktun, we finally obtain 13 baktuns in the baktuns entry.

	12.	19.	19.	17.	19			
+	0.	0.	0.	0.	1			
	12.	19.	19.	17.	20	20 kins	=	1 uinal
	12.	19.	19.	18.	0	18 uinals	=	1 tun
	12.	19.	20.	0.	0	20 tuns	=	1 katun
	12.	20.	0.	0.	0	20 katuns	=	1 baktun
	13.	0.	0.	0.	0			

The answer 13. 0. 0. 0. 0 represents the beginning of a new Maya era of $13 \times 144,000 = 1,872,000$ kins.

Example 5.1.2
The Round Calendar date 8 Ahau 13 Pop corresponding to the Long Count date 9.8.9.13.0 (9 baktuns, 8 katuns, 9 tuns, 13 uinals, 0 kins) is carved in the West panel of the Temple of the Inscriptions (see Figure 5.19). The date represents the birth of the most notable Palenque's king, Lord Pacal, the longest reigning ruler in known Maya history. Lord Pacal was crowned in 9. 9. 2. 4. 8

Figure 5.16 (Schele, 1999) depicts a hieroglyph that reads "It was 12 days, 13 uinals, 17 tuns since he was seated as Lord of the succession Mah K' ina Pacal ahau of Palenque [until] the end of the tun 1 Ahau 8 Kayab the seating of the 10th katun the half-period of the baktun".

The sentence "It was 12 days, 13 uinals, 17 tuns..." is not describing a Long Count date but a time interval between two events called a Distance Number. Notice that in the glyph the Distance Number is written as a reverse form of a Long Count date, see Figure 5.16.

Here we can find the date of the event by adding 12 days(kins), 13 uinals, 17 tuns, in Long Count form 0. 0. 17. 13. 12, to Pacal's accession date 9. 9. 2. 4. 8

	9.	9.	2.	4.	8	Pacal' s accession date		
+	0.	0.	17.	13.	12	Distance number		
	9.	9.	19.	17.	20	20 kins	=	1 uinal
	9.	9.	19.	17 + 1	0			
	9.	9.	19.	18	0	18 uinals	=	1 tun
	9.	9.	19 + 1	0.	0			
	9.	9.	20	0.	0	20 tuns	=	1 katun
	9.	10.	0	0.	0			

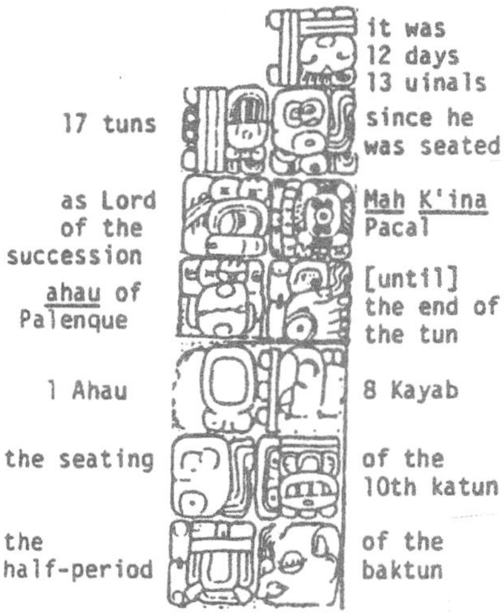

Drawing by Linda Schele © David Schele

Figure 5.16: Glyph (Schele, 1999)

Hence, the event took place in the Long Count date 9. 10. 0. 0. 0 which is at the beginning of the 10th katun since there are no tuns, uinals and kins. The 10th katun is also the half-period of the baktun since 10 katuns = 72,000 kins = $\frac{1}{2} \times 144{,}000$ kins = 1/2 baktun. The Round Calendar date for the event was 1 Ahau 8 Kayab (see exercise 10). The end of the katun was considered a good date for important events.

Example 5.1.3

The Long Count date of Pacal' s accession is 9. 9. 2. 4. 8. We can compute the difference between these two dates to find the ruler's age at the time of his coronation.

First we subtract the corresponding entries obtaining, in this case, two negative entries. To fix this, we have to trade: Since 1 tun = 18 uinals, we borrow 1 tun from the tuns entry (now we have −8 tuns) and trade it as 18 uinals in the uinals entry obtaining 9 uinals. Now we must trade the katun we have for 20 tuns and therefore we obtain 12 tuns and no katuns.

	9.	9.	2.	4.	8	Pacal' s accession date
−	9.	8.	9.	13.	0	
	0.	1.	−7.	−9.	8	
	0.	1.	−8	18 + (− 9)	8	1 tun = 18 uinals
	0.	1.	−8	9.	8	
	0.	0.	20 + (−8)	9.	8	1 katun = 20 tuns
	0.	0.	12.	9.	8	

The answer, in Long Count is 0. 0. 12. 9. 8 which corresponds to $12 \times 360 + 9 \times 20 + 8 = 4,508$ kins. This means that the king was 4,508 days old at the time of crowning and to find his age we divide this number by 365.2425 (the number of days in 1 Gregorian year) obtaining $4,508 \div 365.2425 = 12.34$ years approximately. Consequently, Pacal was 12 years old at the time of accession.

Maya Metonic Cycle

We need to define a few astronomical terms prior to the discussion (Aveni, 1980).

Equinox: Is the point on the celestial sphere at which the sun crosses the celestial equator.

Vernal equinox: Is the point of intersection of the ecliptic and the celestial equator where the sun passes from the southern to the northern hemisphere.

Tropical year: Is a period of revolution of the earth about the sun (or, as we see it, of the sun about the earth) with respect to the vernal equinox; 365.2422 days.

Metonic Cycle: The period (6,939.6 days or 19 years) which returns the full moon to the same date in the calendar year.

In Stela A at Copan, Honduras, we can see the Long Count dates 9. 15. 0. 0. 0, 4 Ahau 13 Yax and 9. 14. 19. 5. 0, 4 Ahau 18 Muan. The difference between the two dates is 0. 0. 19. 5. 0. This Distance number is $19 \times 360 + 5 \times 20 = 6,940$ a number known as the Metonic Cycle connecting the moon cycle and the tropical year.

	9.	15.	19.	5.	0.	4 Ahau	18 Muan
−	9.	15.	0.	0.	0	4 Ahau	13 Yax
	0.	0.	19.	5.	0		

19 years is approximately 235 moon cycles of 29.5 days. This means that if we had a new moon on July 20, 1969 then in 19 more years we have a new moon on July 20, 1988. (Aveni, 1980; Teeple, 1931).

Notice how accurate the Maya calculation of 6,940 days was compared to today's number.

$$19 \text{ tropical years} = 19 \times 365.2422 = 6,939.6 \text{ days}$$

The Distance number 6, 940 days, or 0. 0. 19. 5. 0 can be found in several Maya inscriptions between important Long Count dates (Figure 5.17).

According to Maya mythology, the creation story records the beginning date as the Round Calendar date 4 Ahau 8 Cumku or in Long Count date 0. 0. 0. 0. 0, for some scholars 13. 0. 0. 0. 0. It is believed that a creation has a lenght of 13 baktuns, and therefore the previous creation ended at the conclusion of the 13th baktun, or in Long Count date, the day after 12. 19. 19. 17. 19, December 20, 2012, 3 Cauac 2 Kankin. Another 12. 19. 19. 17. 19 (or 25. 19. 19. 17. 19) will occur on 3 Cauac 17 Chen, followed by 0. 0. 0. 0. 0, (or 26. 0. 0. 0. 0) on May 3, 7138, 4 Ahau 18 Chen. Some scholars believe that at the end of the 13th baktun the Long Count should not reset to 0 baktun, see Table 5.5. Thompson's theory (1960) is that the baktun cycle is a 20-baktun cycle.

A rare long count date can be seen in Stela 1 located in Coba, Yucatan, Mexico, which has 24 entries

13.13.13.13.13.13.13.13.13.13.13.13.13.13.13.13.13.13.13.13.0.0.0.0 in the long count (Stuart, 2011) (Figure 5.18).

Figure 5.17: Stela A

Table 5.5:

Reseting	Not reseting	Round Calendar	Gregorian	Creation
0. 0. 0. 0. 0 + 1 day	13. 0. 0. 0. 0	4 Ahau 8 Cumku	Aug. 11, 3114BC	End of a creation
0. 0. 0. 0. 1 + 13 baktuns − 2 days	13. 0. 0. 0. 1	5 Imix 9 Cumku	Aug. 12, 3114BC	First day of a new creation
12. 19. 19. 17. 19 + 1 day	25.19.19.17.19	3 Kawak 2 Kankin	Dec. 20, 2012AD	Day before the end of a creation
13. 0. 0. 0. 0 + 1 day	26. 0. 0. 0. 0	4 Ahau 3 Kankin	Dec. 21, 2012AD	End of a creation
0. 0. 0. 0. 1 + 13 baktuns	26. 0. 0. 0. 1	5 Imix 4 Kankin	Dec. 22, 2012AD	First day of a new creation
0. 0. 0. 0. 0 + 1 day	39. 0. 0. 0. 0	4 Ahau 18 Chen	May 3, 7138AD	End of a creation
0. 0. 0. 0. 1	39. 0. 0. 0. 1	5 Imix 19 Chen	May 4, 7138AD	First day of a new creation

Courtesy of Ximena Catepillán

Figure 5.18: Stela 1 at Coba

The example at Coba, on Stela 1, places nineteen 13's above the end of a creation date: 13. 0. 0. 0. 0 August 11, 3114 BC. If we attempted to actually count all those 13's as actual coefficients, each going up by a magnitude of 20, the date would go 28,285,978,483,664,581,446,157,328,238,631 years into the past (Stuart, 2011).

In another inscription, this one from Yaxchilan, a scribe wrote the date 13. 13. 13. 13. 13. 13. 13. 13. 9. 15. 13. 6. 9 corresponding to 3 Muluc 17 Mac, on the stairs of Temple 33.

So, perhaps the idea that the 13th Baktun is the end of an era is just symbolic, we experienced one of the endings in December 2012. Maya texts say nothing about what might happen at the 13th Baktun, but we do know that the end of each baktun was considered a time of great change. Reviewing ancient Maya history provides a general corroboration of that notion (Maya Exploration Center).

CALENDAR CONVERSIONS:

In this section, we present a method of converting a Long Count date to the corresponding Round Calendar date. This means, given a Long Count date, we locate it in the Round Calendar 18,980-day cycle by identifying the corresponding Tzolkin and Haab dates.

Example 5.1.4

The Round Calendar date 8 Ahau 13 Pop corresponding to the Long Count date 9.8.9.13.0 (9 baktuns, 8 katuns, 9 tuns, 13 uinals, 0 kins) is carved in the West panel of the Temple of the Inscriptions in Palenque, Mexico (see Figure 5.19). The Gregorian year was 603 AD and the date represents the birth of the most notable Palenque's king, Lord Pakal. Pakal was the longest reigning ruler in known Maya history, ruling 68 years (615AD–683AD), from the age of 12 until 80. Lord Pacal accession date is recorded below in the same panel, 5 Lamat 1 Mol a Round calendar date that corresponds to the Long Count date 9. 9. 2. 4. 8.

We show how to convert these two important Long Count dates in Pacal's life into the Round calendar dates that can be seen in the hieroglyphic panel.

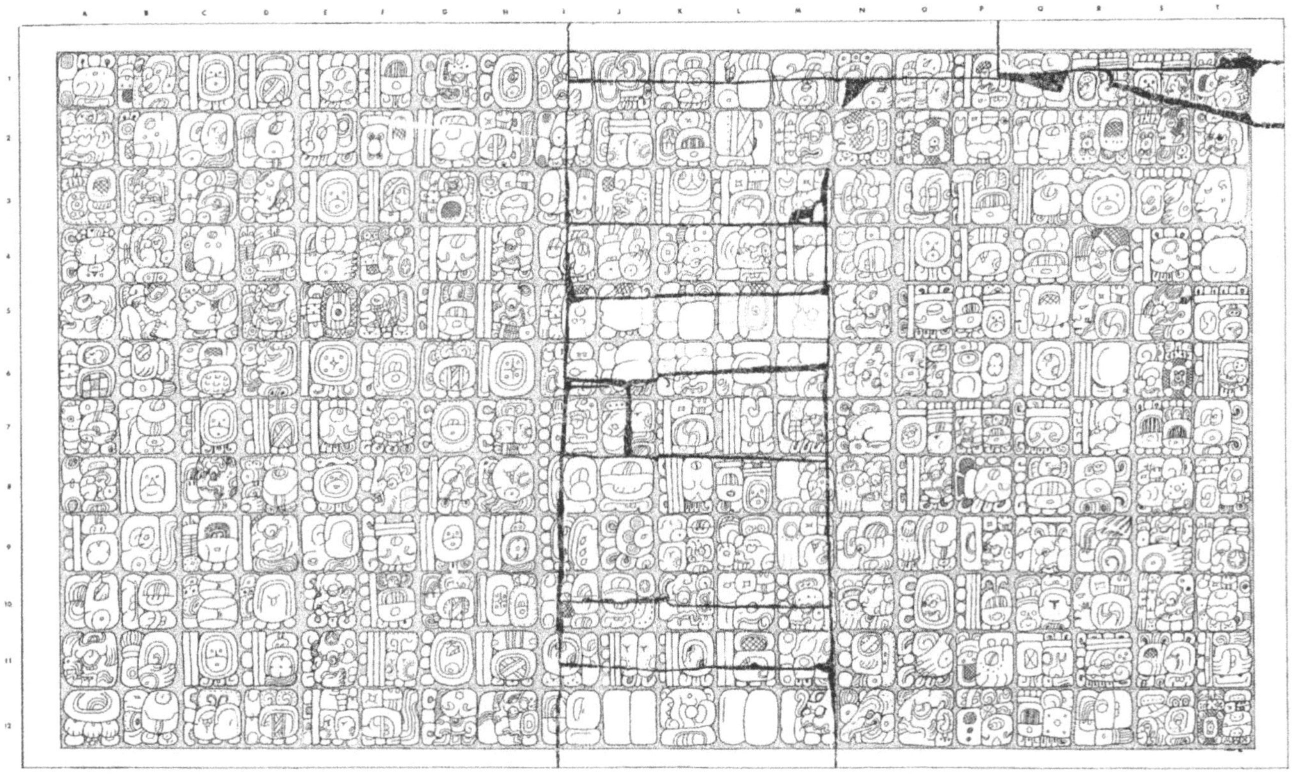

Drawing by Linda Schele © David Schele

Figure 5.19: West Panel Temple of the Inscriptions (Schele, 1999)

To find the Tzolkin date:

Step 1:
First we compute the number of kins.

Baktuns	9 × 144,000	=	1,296,000
Katuns	8 × 7,200	=	57,600
Tuns	9 × 360	=	3,240
Uinals	13 × 20	=	260
Kins			0
	Total	=	1,357,100

In what follows we use the Division Algorithm several times.

Step 2:
Now we divide the number of kins by 18,980 to determine the number of full Round Calendar cycles that elapsed up to this number of kins, and, more importantly, to find the remainder, which gives the location of the date within one Round Calendar cycle:

$1,357,100/18,980 = 71 + 9,520/18,980.$

This means that 71 Round Calendar cycles elapsed and the date we seek is the 9,520th kin or day from the starting date of 4 Ahau 8 Cumku in the Round Calendar cycle. The remainder 9,520 is the key to find the date.

Step 3:
To find the Tzolkin date we divide the remainder 9,520 by 13 (days) and by 20 (gods):

$9,520/13 = 732 + 4/13$, remainder 4

$9,520/20 = 476 + 0/20$, remainder 0

Therefore, 9,520 days consist of 732 13-day cycles, plus 4 extra days, and also of 476 20-day cycles of gods, plus 0 extra days (or gods). Since the Tzolkin date for the beginning of the Maya world was 4 Ahau, we add 4 days to 4, and 0 gods to Ahau, obtaining the Tzolkin date 8 Ahau.

To find the Haab date:

Step 1:
We begin by dividing the remainder 9520 by 365, the number of days in the Haab calendar:

$9520/365 = 26 + 30/365.$

This means that 9520 days consist of 26 Haab Calendar cycles, plus 30 extra days. Since the Haab date for the beginning of the Maya world was 8 Cumku, we add 30 days to 8 days, obtaining 38 days, the Haab Calendar date is 13 Pop. Because 30 days after 8 Cumku is 13 Pop. See list of days below:

8 Cumku

9 Cumku

...

19 Cumku (11 days)

0 Uayeb

1 Uayeb

...

4 Uayeb (5 days)

0 Pop

1 Pop

...

13 Pop (14 days)

Thus, the Long Count date corresponds to 8 Ahau 13 Pop.

The method utilized for the Haab date computation can be summarized as follows: (Catepillán and Szymanski, 2010)

Note that if the remainder R satisfies $0 \leq R \leq 11$, the Haab Calendar date is in the month of Cumku and is 8 + R Cumku. If $12 \leq R \leq 16$, the Haab date is one of the 5 days of Uayeb; namely, R – 12 Uayeb. Finally, if $17 \leq R \leq 364$, the Haab Calendar date falls between 0 Pop and 7 Cumku (inclusive).

The calendar converter developed by the Maya Exploration Center converts Gregorian calendar dates to Long Count, Tzolkin calendar, and Haab calendar dates. In addition, a similar converter developed by the Smithsonian National Museum of the American Indian includes drawings depicting how the Maya utilized glyphs to write the dates. The reader should use these converters to check the accuracy of the calendar conversions they perform by hand.

Example 5.1.5
Let's convert to Round Calendar date the Long Count date 9. 9. 2. 4. 8, the accession date of Lord Pacal.

To find the Tzolkin date:

Step 1:
First we compute the number of kins.

Baktuns	9 × 144,000	=	1,296,000
Katuns	9× 7,200	=	64,800
Tuns	2 × 360	=	720
Uinals	4 × 20	=	80
Kins			8
Total		=	1,361,608

Step 2:
Now we divide the number of kins by 18,980 to determine the number of full Round Calendar cycles that elapsed up to this number of kins, and, more importantly, to find the remainder, which gives the location of the date within one Round Calendar cycle:

$1,361,608/18,980 = 71 + 14,028/18,980.$

This means that 71 Round Calendar cycles elapsed and the date we seek is the 14,028th kin or day from the starting date of 4 Ahau 8 Cumku in the Round Calendar cycle. The remainder 14,028 is the key to find the date.

Step 3:
To find the Tzolkin date we divide the remainder 14,028 by 13 (days) and by 20 (gods):

$14,028/13 = 1,079 + 1/13$, remainder 1

14,028/20 =701 + 8/20, remainder 8

Therefore, 14,028 days consist of 1,079 13-day cycles, plus 1 extra days, and also of 701 20-day cycles of gods, plus 8 extra days (or gods). Since the Tzolkin date for the beginning of the Maya world was 4 Ahau, we add 1 day to 4 days and 8 gods to Ahau, obtaining the Tzolkin date 5 Lamat.

To find the Haab date:

Step 1:
We begin by dividing the remainder 14,028 by 365, the number of days in the Haab calendar:

$14,028/365 = 38 + 158/365.$

This means that 14,028 days consist of 38 Haab Calendar cycles, plus 158 extra days. Since the remainder 158 satisfies $17 \leq 158 \leq 364$, the Haab date falls between 0 Pop and 7 Cumku (inclusive).

The number of days we count from 8 Cumku is then $158 - 16 = 142$ which corresponds to 7 Haab months of 20 days plus 2 days. Thus, the date fall on the second day of the eighth month, i.e., 1 Mol.

Since the Haab date for the beginning of the Maya world was 8 Cumku, we add 158 days to 8 days, the Haab Calendar date is 1 Mol, because 158 days after 8 Cumku is 1 Mol. See list of days below:

8 Cumku

9 Cumku

...

19 Cumku (11 days)

0 Uayeb

1 Uayeb

...

4 Uayeb (5 days)

0 Pop

1 Pop

...

19 Pop (20 days)

At this point we have 36 days have elapsed since 8 Cumku and we still need to add 122 more days, i.e., 6 months of 20 days each plus 2 more days. Therefore we end at 1 Mol.

Thus, the Long Count date corresponds to 5 Lamat 1 Mol.

In Summary
The method described in Examples 1 and 2 can be expressed in general terms as follow: (Catepillán and Szymanski, 2010)

Compute the number of kins K.

Divide K by 18,980; the remainder is R.

Tzolkin date: Divide R by 13 and by 20 to obtain respective remainders A and B. To get the Tzolkin number, we imagine the numbers 1 through 13 on a circle, locate 4 and move A places ahead. To get the Tzolkin god, we look at the list of Tzolkin gods and, imagining them on a circle, locate Ahau and move B places ahead.

Haab date: Divide R by 365; the remainder is H.

If $0 \leq H \leq 11$, the Haab date is $8 + H$ Cumku.

If $12 \leq H \leq 16$, the Haab date is H – 12 Uayeb.

Finally, if $17 \leq H \leq 364$, the Haab date falls between 0 Pop and 7 Cumku (inclusive), and can be found as follows.

Compute H' = H – 17.

Divide H' by 20; the quotient is Q, the remainder is W.

The Haab number is W and the Haab month (god) is Pop + Q, so that the Haab date is W Pop + Q.

Finally, the Round Calendar date is the Tzolkin date followed by the Haab date.

How to convert a Long Count date to Julian Calendar: (Schele, page 167, in Coe and Van Stone, 2001)

Step 1:
Find the number of kins K on the Long Count date.

Step 2:
Add 584,285 (a correlation constant) to the number of kins:

K + 584,285

Divide by 365.25 (the number of days in a Julian Calendar year):

(K + 584.285) ÷ 365.25 = L

Step 3:
Find L – 4,712. The integer part of this number is the Julian Calendar year.

Step 4:
Multiply the decimal part of L – 4712 by 365.25 and round upward. Let's call the number M.

Step 5:
The number M obtained in step 4 satisfies $1 \leq M \leq 366$. Now use Table 5.6 below to find the nearest number $N \leq M$. The Julian month is on the left by the number and the day of the month is M – N

The Julian Calendar date is:

Day = M – N

Month = Written on the table by N

Year= Integer part of L – 4,712

Example:
We find the Julian Calendar date corresponding to the Long Count date 8. 12. 14. 8. 15 found in Stela 29 in Tikal, Guatemala.

Table 5.6: To convert Long Count date to Julian Calendar date

January	0	May	120	September	243
February	31	June	151	October	273
March	59	July	181	November	304
April	90	August	212	December	334

Step 1:
The number of kins K = 1,243,615 can be found above Figure 5.15.

Step 2:
$K + 584,285 = 1,243,615 + 584,285 = 1,827,900$

$(K + 584.285) \div 365.25 = 1,827,900 \div 365.25 = 5,004.517 = L$

Step 3:
$L - 4,712 = 5,004.517 - 4,712 = 292.517$ The integer part of this number, 292, is the Julian Calendar year.

Step 4:
Multiply the decimal part of $L - 4712$ by 365.25.

Decimal part of $(L - 4712) = 0.517$.

$0.517 \times 365.25 = 188.834$ and rounding upward we obtain $189 = M$

Step 5:
$M = 189$ in step 4 satisfies $1 \leq M \leq 366$. Now use the table below to find $N = 181$, the nearest number $N \leq M$. The Julian month July is on the left of 181 and the day of the month is $M - N = 189 - 181 = 8$

The Julian Calendar date is July 8, 292 AD.

How to calculate the Lord of the Night on a Long Count date:
The Long Count date 12. 19. 15. 15. 8 in Figure 5.13 has a G2 Lord of the Night. As we mentioned before there are 9 Lords of the Night labeled G1, G2, ..., G9 and to find the number attached to the G we follow a simple process; find the number of kins in the last two entries of the Long Count date, in this case $15 \times 20 + 8 = 308$ and then calculate the remainder after dividing the number by 9. The remainder is the number by the G Lord of the Night.

Therefore, $308 \div 9 = 34.22$ and $308 - 34 \times 9 = 2$, thus, the remainder is 2 and therefore, the Lord of the Night is G2. Thus, if G2 is the Lord of the Night for the Long Count date 12. 19. 15. 15. 8, then we can find the Lord of the Nights for days before and after the given Long Count date using a 9-hour clock.

Note: When dividing by 9 we never get remainder 9, hence, we assign Lord of the Night G9 everytime we have a remaider 0 (Figure 5.20).

Illustrations by John Montgomery from *Breaking the Maya Code* by Michael D. Coe, Thames & Hudson, London and New York.

Figure 5.20: The Nine Lords of the Night (Coe, 1992)

Example:
What is the Lord of the Night for the Long Count date 12.19. 15. 16. 4? Here the difference between the dates is $(16 \times 20 + 4) - (15 \times 20 + 8) = 16$, the need to move 16 gods forward starting with god G2. Simply move 16 hours from hour 2 in the 9-hour clock, the number is 9, hence, the god is G9.

What matters in the computation is only the last 2 entries in the Long Count date, therefore, the Lord of the Night for a Long Count date x. y. z. 15. 8 is also G2.

How to approximate a Long Count date to Gregorian date:
In example 5.1.4 we converted the Long Count date 9. 8. 9. 13. 0 carved in the Temple of the Inscriptions, a date that represents the birth of the most notable Palenque's king, Lord Pacal. In the first conversion step we found the number of kins to be 1,357,100. To find the Gregorian date we divide the number of kins by 365.2425 (1 Gregorian year which corresponds to 365 days, 5 hr, 49 min, and 12 s) and obtain:

$$1,357,100 \div 365.2425 = 3,715.6136 \text{ years}$$

Now we add the year of the beginning of the Maya era 3114 BC (we are substracting 3,114 because it is a BC year) to this year and obtain 601.6136, therefore the approximate year is 602 AD.

EXERCISES

1. The earliest Maya Long Count date yet discovered was carved on Stela 2 at Chiapa de Corzo (in modern day Chiapas). It records 7.16.3.2.13, or December 36 BC. Find the number of kins for the Long Count date.
2. Find the Tzolkin Calendar and Haab Calendar dates for the Long Count date in Exercise 1.
3. Find an approximation to the Gregorian year for the Long Count date in Exercise 1.
4. Find the Long Count date 10,000 days after the Long Count date given in Exercise 1.
5. **a.** Find the Long Count date for the first day of the 62nd cycle of 18,980 in the Round Calendar.
   b. Show that the Round Calendar date for the Long Count date is 4 Ahau 8 Cumku.
6. If the fifth cycle number of kins is 94,900, find the number of kins for the first day of the 12th cycle.
7. On the Long Count date 9. 8. 19. 7. 18 Lady Zac-Kuk, Pacal's mother acceded at Palenque.
   a. Find the Round Calendar date.
   b. Find the Round Calendar date 15 days later.
   c. Find the Round Calendar date 15 days before.
8. In Tikal, Guatemala, the oldest Long Count date found in the area is 8. 12. 14. 8. 15 (Stela 29) and the most recent is 10. 2. 0. 0. 0 (Stela 11).
   a. Find the distance number between the dates.
   b. Determine approximately the number of years between the dates. This computation will give you an approximation of the duration of the Maya settlement at Tikal.
9. Find the Long Count dates corresponding to the following number of kins.
   a. 1,302,884 **b.** 1,383,136 **c.** 1,872,632
10. Show that the Round Calendar date for the Long Count date 9. 10. 0. 0. 0 was 1 Ahau 8 Kayab.
11. Explain why Long Count dates of the form a. b. 0. 0. 0, $a \neq 0$ and $b \neq 0$, have a Tzolkin god Ahau.
12. Convert the Long Count dates 8. 12. 14. 8. 15, and 10. 2. 0. 0. 0 from Exercise 8 to Round Calendar dates.
13. **a.** Use Maya Exploration Center' s calendar conversion site to find the Long Count date of your birthday.
    b. Find the Round Calendar date corresponding to your birthday and check the answer against the date on the site.
14. The Maya predicted the total solar eclipse that took place July 11, 1991 (Bricker et al., 1983). Find the Long Count date for the eclipse and convert the date into Round Calendar date.
15. The Long Count date 9. 9. 16. 0. 0 was found on page 24 of the Dresden Codex. Explain how the number of kins on this date is related to the number of kins in the Tzolkin, Haab and Round Calendars.

16. Find the Julian Calendar dates for the Long Count dates found in Exercise 9.

17. Find the Lord of the Night, GX, $1 \leq X \leq 9$, for the Long Count dates found in Exercise 9.

18. Explain why all Long Count dates that are tun endings have a G9 Lord of the Night.

19. If the Lord of the Night for the Long Count date 9. 9. 11. 12. 7 is G4 find the Lord of the Night for the Long Count date 9. 9. 9. 12. 0.

REFERENCES

Aveni, Anthony. *Skywatchers of Ancient Mexico*. University of Texas Press, Austin, 1980.

Bohm, V., Bohm, B., Klokocnik, J., Vondrak, J., and Kostelecky, J. Dating of Mayan Calendar Using Long-Periodic Astronomical Phenomena in Dresden Codex, *Serbian Astronomical Journal*, 186, 53–64, 2013.

Bricker, Harvey; Bricker, Victoria; Aveni, Anthony; Closs, Michael; Edmonson, Munro; Lounsbury, Floyd; and Taladoire, Eric. Classic Maya Prediction of Solar Eclipses, *Current Anthropology*, Vol. 24, No. 1, 1983.

Caso, Alfonso. *Los Calendarios Prehispanicos*, Universidad Nacional Autonoma de Mexico, Mexico, 1967.

Catepillán, Ximena and Szymanski, Waclaw. *Maya Calendar Conversions, MAA Convergence*, 2010. http://www.maa.org/press/periodicals/convergence/maya-calendar-conversions

Coe, Michael. *Breaking the Maya Code*, Thames & Hudson, New York, 1992.

Coe, Michael and Van Stone, Mark. *Reading the Maya Glyphs*, Thames & Hudson Ltd, London, 2001.

Coe, William R. *Tikal*: *Guía de las Antiguas Ruinas Mayas*, Piedra Santa, Guatemala, 1967.

De Landa, Diego. *Relación de Las Cosas de Yucatán*, Editorial San Fernando, Mexico, 2005 (Version original 1566).

Edmonson, Munro S. *Sistemas Calendaricos Mesoamericanos*, Universidad Nacional Autonoma de Mexico, Mexico, 1995.

Gendrop, Paul. *Arte Prehispánico en Mesoamerica*, Editorial Trillas, Mexico City, 1970.

Houston, S., Chinchilla, O., Stuart, D. *The Decipherment of Ancient Maya Writing*, University of Oklahoma Press, Norman, OK, 2001.

Lounsbury, Floyd. Maya Numeration, Computation, and Calendrical Astronomy, *Complete Dictionary of Scientific Biography*, 1978.

Vol. 15, Charles Scribner's Sons, Detroit, 2008. pp. 759–818. COPYRIGHT 2008 Charles Scribner's Sons, a part of Gale, Cengage Learning.

Morley, Sylvanius, George Brainerd, and George Sharer. *La Civilización Maya*, Fondo de Cultura Economica, Mexico, 1947a.

Morley, Sylvanus, Brainerd, George, and Sharer, George. *The Ancient Maya*, Stanford University Press, Stanford, CA, 1947b.

Schele, Linda. *A Palenque Triad*, Maya Workshop Foundation, Austin, Texas, 1999.

Stuart, David. *The Order of Days*, Harmony Books, New York, 2011

Teeple, John E. *Maya Astronomy, Contributions to American Archaeology*, Vol. I, No. 2, Carnegie Institution of Washington, Washington, 1931.

Thompson, J. Eric. *Maya Hieroglyphic Writing: An Introduction*, University of Oklahoma Press, Norman, 1960.

Tozzer, A., *Landa's Relación de Cosas de Yucatan, Cambridge: Papers of the Peabody Museum of American Archaeology and Ethnology*, Vol. 18, Harvard University, Cambridge, 1941.

Wagner, Elizabeth. *Maya Creation Myths and Cosmography*, Maya, Divine Kings of the Rain Forest, Könemann, 2001.

Resources

Maya Exploration Center: http://www.mayaexploration.com/index.php

Maya Exploration Center: http://maya.nmai.si.edu/calendar/maya-calendar-converter

Pictographic Maya Calendar Converter:

Hartley (2002) in http://users.hartwick.edu/hartleyc/mayacalendar/mayacalendar.html

http://mathdl.maa.org/mathDL/46/?pa=content&sa=viewDocument&nodeId=3886

Figure 5.1: http://www.famsi.org/mayawriting/codices/madrid.html

Figure 5.3: Schele (2000) in http://www.latinamericanstudies.org/chiapa-de-corzo.htm

www.ingramcontent.com/pod-product-compliance
Ingram Content Group UK Ltd.
Pitfield, Milton Keynes, MK11 3LW, UK
UKHW050722170926
15229UKWH00010B/42